Mine Pumping Engines

in Eighteenth Century Cornwall

R. J. Stewart

Published by The Trevithick Society
for the study of Cornish industrial archaeology and history

The Trevithick Society is a Charitable Incorporated Organisation
Registered Charity no. 1159639

ISBN: 978-0-9935021-2-5

Printed and bound by Short Run Press Ltd.,
25 Bittern Rd, Exeter EX2 7LW

Typeset by Peninsula Projects
c/o PO Box 62, Camborne, Cornwall TR14 7ZN

Statue of Watt, Boulton and Murdock in central Birmingham. It is highly unlikely, given the local antipathy to Boulton and Watt, that a similar statue could have been erected in Cornwall.

This book is dedicated to my partner Kim who looked after me beautifully when I was broken and who puts up with my many and varied idiosyncrasies.

Author's Foreword

I had no intention of writing this book.

Its roots lie in my long term interest in the Coster family. As someone who reads voraciously, I kept coming across the name Coster in all its myriad spellings. A number of authors made tantalising allusions to John Coster / Costar / Kostar and his role in the development of Cornish copper mining. Such information as there was, tended to be intriguingly contradictory and ephemeral. My interest in the Costers developed into a project which I tinkered with as a break from researching my book on Devon Great Consols, published in 2013. It culminated in an article about the Costers published in the 2015 Journal of the Trevithick Society.

The more I looked into the activities of the Costers, the more I developed an interest in eighteenth century technology, particularly the Newcomen engine. Unlike the Costers the Newcomen engine has a vast and growing literature, albeit little relating directly to Cornwall. As an adjunct to my work on the Costers, I started compiling notes on Newcomen engines in Cornwall and by early 2015 I had the makings of an article on the subject. However by this time my interest had shifted and I had started work on a history of the mines of Calstock Parish, the Newcomen material being put on the back burner.

Fast forward to Christmas 2015 and rather foolishly I managed to rupture my Achilles tendon. Trying to find a positive angle on the situation I rationalised that this meant a lot of time to devote to research and writing. A ruptured Achilles tendon meant that I could not walk, let alone drive; which in turn meant that I could not get to libraries and archives. This pretty much scuppered any chance of working on the Calstock Parish project which draws heavily on the *Mining Journal*, the nearest set being in the care of the Cornish Studies Library at Redruth. I then started to look around for a project that I could complete without the need to travel (or indeed move too far from my desk).

It was at this point that the sidelined Newcomen engine article sprang to mind, and it struck me that it would be an interesting exercise to widen the scope of the project to cover the whole of the eighteenth century up to the expiry of Watt's patent in 1800. I am fortunate to have built up a fairly extensive personal library, this, supplemented by Google books and the availability of online transcriptions of Boulton and Watt's correspondence with Thomas Wilson convinced me that this project was a practical proposition. Given the constraints I have been working under, I have drawn more heavily than I would have liked on previous authors' published work. This at times may make this work feel derivative, for which I can only apologise. However I do hope that on occasion I have made some original contributions to the subject.

If nothing else this project has saved me from the delights of daytime TV!

Rick Stewart, Calstock, February 2017.

Acknowledgements.
Aditnow, Birmingham Library (Wolfson Centre), Clive Briscoe, Tony Brooks, Tony Clarke, Ainsley Cocks, Cornish Studies Library (Redruth), Cornwall Record Office (Truro), Cornwall and West Devon Mining Landscape World Heritage Site Office, Lionel Ford, Hugo Glazer, James Greener, Steve Grudgings, Ironbridge Gorge Museum Trust, Andrew Hutcheon, Pete Joseph, Chris Kelland, King Edward Mine, Michael Messenger, The Mining History List, Morrab Library (Penzance), Royal Institution of Cornwall Library (Truro), Graham Thorne, Professor Hugh Torrens, Dave Warne, Robert Waterhouse.

Several images from the Boulton and Watt Collection held by the Library of Birmingham have been used in this publication. The original drawings are well over 200 years old and are in many cases very fragile, faded and in some cases damaged. In consequence they do not reproduce well photographically, with much detail being lost. In spite of that the drawings are of such technical and historical interest that they have been included in this publication with all their imperfections.

The author and publisher are grateful for the generous help and time given by the staff at the Wolfson Centre in locating and producing the Boulton and Watt drawings.

World Heritage Site Foreword

The Cornwall and West Devon Mining Landscape World Heritage Site Partnership is pleased to support the publication of *Mine Pumping Engines in Eighteenth Century Cornwall.*

Cornwall has been described by Emeritus Professor of Mining History Roger Burt* as being '...probably the most important mining district in the world...' by the 1850s. Certainly its prodigious output of copper and tin was to be strongly felt on world metal markets for much of the nineteenth century. A legacy of this activity is an outstanding mining landscape populated by numerous Cornish type engine houses – the most instantly recognisable feature of the Cornwall and West Devon Mining Landscape World Heritage Site.

The position of dominance symbolised by these icons of industry could not have been secured without the development of steam pumping. While the use of water wheels for mine pumping had been established in the south west by the late fifteenth century - and the efficiency of these later greatly improved by John Coster II - the development of deep lode copper mining in Cornwall and Devon would not have been possible if not for the progressive application of steam technology. The eighteenth century steam engine designs of Newcomen, Smeaton, Watt and others were to render possible the growth of Cornwall to a position of copper mining pre-eminence.

While the progress of the high-pressure Cornish steam engine is generally well addressed within the annals of Cornish metalliferous mining, what preceded it is less well understood. This work by Rick Stewart draws together a wide range of bibliographic sources relating to early mine pumping in Britain and, in doing so, provides a narrative of how incremental changes in steam technology were to set the scene, ultimately in the nineteenth century, for the Cornish engine. This addresses a significant gap in the literature of Cornish mining and is to be much welcomed by anyone with an interest in how Cornwall was to become, for a time, a global mining 'giant'.

Julian German

Chair , Cornwall and West Devon Mining Landscape World Heritage Site Partnership

January 2017

www.cornishmining.org.uk

*Burt, R., Burnley, R., Gill, M. and Neill, A. (2014) *Mining in Cornwall & Devon - Mines and Men.* Exeter: University of Exeter Press, p.15

Publisher's Note

The early years of mine pumping in Cornwall have received far less attention than those of the Cornish engine when Cornwall was briefly at the forefront of technological innovation. The eighteenth century period was thus overdue for re-examination. This book's publication coincides with the first International Early Engines Conference at Elsecar, Yorkshire in May 2017 confirming that considerable research in recent years has concentrated on the birth pangs of mechanical pumping.

When Rick Stewart, author of the Trevithick Society's bestselling history of Devon Great Consols and a participant in the IEE Conference, brought this project to us, we were immediately keen to publish his work which brings together much of the recent research to fill a gap in Cornish mining literature. The Society has been hugely assisted and encouraged by the support of the Cornish Mining World Heritage Site Office in bringing this important work to fruition and we gladly acknowledge this. Although there is much still to discover about these early years, we believe that *Mine Pumping Engines in Eighteenth Century Cornwall* will remain a key text for the foreseeable future.

Introduction

When one thinks of mining in Cornwall, what comes to mind are the ruined engine houses which pepper the landscape; they are the enduring and, indeed iconic image of the county. These are, almost without exception, nineteenth century in origin. They housed engines worked by high pressure steam, Cornwall's legacy to the world. The engines these houses contained were designed by the great Cornish engineers of the day; names such as Trevithick, Woolf, Sims, Loam and West are familiar to all but the most casual student of Cornish industrial history. The engines were built by the great Cornish engineering firms such as Harveys of Hayle and the Copperhouse and Perran Foundries. The "Cornish" pumping engine reached a pinnacle of excellence during this period; West's 80″ at Fowey Consols, arguably being *primus inter pares*. Few would question that for a time during the nineteenth century Cornish engineers and engineering led the world. To many Cornishmen this achievement is rightly a great source of pride and so has generated a considerable literature. Whilst the development of the Cornish engine during the nineteenth century is very well documented in contrast its antecedents in the preceding century are less well known and less well understood; hopefully this book goes some way to remedy that situation.

The Cornish pumping engine did not appear as a fully formed entity at the beginning of the nineteenth century, it had a long gestation, the earliest roots of which may be found as far back as the Tudors. It was during the Tudor period that a significant revolution took place in Cornish mining, namely the transition from tin streaming to lode mining. As miners went ever deeper in search of tin ore they increasingly encountered water, and so the question of mine drainage came to the fore. The tinners addressed the problem with considerable ingenuity, utilising short drainage adits, horse whims for bailing water, rag and chain pumps and waterwheel operated lift pumps.

A second revolution took place at the end of the seventeenth and early eighteenth centuries, the copper revolution. This was a change every bit as significant as the Tudor transition from tin streaming to lode mining. The copper revolution was initiated by Clement and Talbot Clerke who by 1689 had successfully developed a method of smelting copper ore with coal. Prior to this the English brass and copper industries had been dependent on imported copper, Sweden being a very important supplier. This technological breakthrough meant that the brass and copper industry which at this time was dominated by Bristol could exploit English sources of copper ores. Previous attempts to develop copper mining in both Devon and Cornwall, most notably in the 1580s under the auspices of the Company of Mines Royal, had been largely unsuccessful.

At the beginning of the 1690s copper mining activity in Cornwall was minimal, albeit starting to develop to meet the burgeoning demand of the Bristol brass and copper industries. The earliest Cornish copper mine of any real significance was the mine at Chacewater which was developing as a copper producer in the early 1690s. To understand the challenge faced by late seventeenth and early eighteenth century copper miners one needs to consider the characteristics of a "typical" copper lode (if such a thing exists). In our typical lode the upper section of the ore body, known as the gossan, is heavily weathered by the percolation of surface water. The gossan is denuded of copper which is leached out of this upper zone and re-deposited as secondary copper minerals at depth. Cassiterite is less susceptible to leaching and tends to remain in the gossan. Below the gossan lies the unweathered copper zone containing both primary and secondary copper minerals. The boundary between the upper tin zone and the lower copper zone is typically marked by the level of the water table. This zoning and the transition from tin to copper was well understood by the pioneer copper miners. Writing in the mid 1720s Henric Kalmeter, a Swedish industrial "spy", comments:

> "....... these mines were taken up and worked for tin, until, when they came further and deeper down the lode, the former diminished or disappeared, and copper come in instead."[1]

The tinners had the distinct advantage that they could extract cassiterite from the comparatively easily worked weathered ground which lay above the water table. In contrast the would-be copper miner was working below the water table in hard, unweathered ground. If copper was to be mined successfully in Devon and Cornwall, the twin challenges of breaking hard ground and drainage at increasing depths had to be addressed. The introduction of gunpowder into the South West in 1689 effectively dealt with the issues surrounding the breaking of hard ground.

As mines grew steadily deeper during the eighteenth century miners and engineers had repeatedly to re-evaluate and reconsider their responses to keeping the mines drained; it is these solutions to the problem which form the subject of this book. The intention is to follow the development of the pumping engine in Cornwall through the eighteenth century starting with an examination of the innovations in water management and the use of water engines. The introduction of Thomas Newcomen's atmospheric engine in the 1710s to its ultimate expression in John Smeaton's magnificent 72″ atmospheric engine erected at Chacewater Mine in the mid 1770s is explored in some detail. There follows an account of Boulton and Watt's activities in the county and finally details of the engines erected by Jonathan Hornblower and Edward Bull in defiance of Watt's 1769 patent.

Unlike the story of the Cornish pumping engine in the nineteenth century which is one of indigenous endeavour, the eighteenth century story is one of how Cornish mining and engineering progressed by adopting technology developed by "outsiders" and "incomers": The Coster family, who kick-started this technological revolution, came from the Forest of Dean whilst their financial backing came from the Bristol brass and copper industries. Thomas Savery and Thomas Newcomen were both Devonians. John Smeaton was a Yorkshireman and James Watt was a Scot, whilst his partner Matthew Boulton was a Midlander. During the eighteenth century somewhere in the region of one hundred and forty engines were erected in Cornwall, practically all of which were manufactured outside the county, most notably at Coalbrookdale. Whilst then this is to a large extent a story of the contribution made by outsiders and incomers, it is also a story of assimilation. The Hornblower family found its way into Cornwall from Coalbrookdale as erectors of Newcomen engines; once there they stayed, spawning a generation of Cornish engineers. Edward Bull came to Cornwall as one of Boulton and Watt's engine erectors but quickly branched out on his own, taking a young Richard Trevithick under his wing and by so doing gave him a sound grounding in engine construction and a taste for unconventional engineering solutions. The proliferation of engines in the county from mid century onwards also allowed indigenous engineers such as Budge and Nancarrow to develop their skills. Cornish engineering developed from almost nothing at the beginning of the eighteenth century to a point in the early nineteenth century where it was ready to conquer the world.

References

1. Brooke J., 2001

Contents

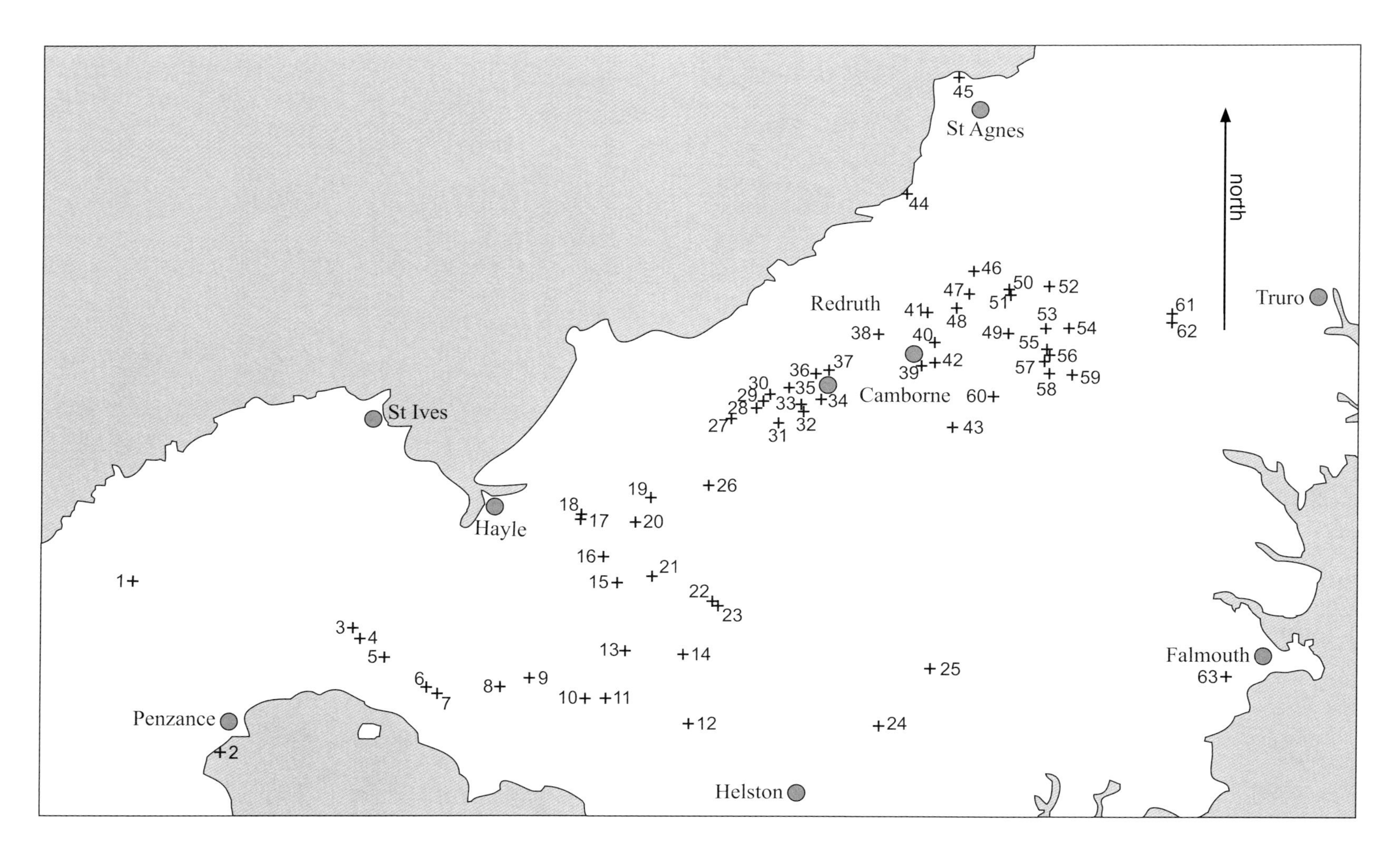

St Agnes
north
Redruth
Truro
Camborne
St Ives
Hayle
Falmouth
Penzance
Helston

List of mine locations in central and west Cornwall.

1. Ding Dong
2. Wherry
3. Ludgvan Lease
4. Wheal Bolton
5. Wheal Fortune (Ludgvan)
6. Tregurtha Downs
7. Owen Vean
8. Hallamaning
9. Wheal Fortune (St Hilary)
10. Wheal Reeth
11. Great Work
12. Wheal Vor
13. Godolphin
14. Rosewarne
15. Carsize
16. Carloose
17. Prince George
18. Herland
19. Trevaskis
20. Relistian
21. Treasure
22. Wheal Abraham
23. Clowance
24. Trevenen
25. Tregonibris
26. Bosprowel
27. Crane
28. Weeth
29. Wheal Chance
30. Wheal Kitty
31. Gons
32. Dolcoath
33. Bullen Garden
34. Tincroft
35. Roskear
36. Trevenson
37. Pool
38. Tolgus
39. Sparnon
40. Cardrew
41. Treleigh Wood
42. Pednandrea
43. Tresavean
44. Towan
45. Pell
46. Wheal Rose (Scorrier)
47. North Downs
48. Wheal Fortune (Scorrier)
49. Gorland
50. Scorrier
51. Wheal Fat
52. Wheal Busy/Pittslooarn
53. Poldice
54. Creegbrawse
55. Wheal Maid
56. Wheal Virgin
57. West Wheal Virgin
58. Poldice
59. Ale & Cakes
60. Ting Tang
61. Baldue
62. Tregothnan
63. Swanpool

Chapter 1
Adits and Water Engines

As outlined in the introduction, the 1690s saw the emergence of deep copper mining in Devon and Cornwall, largely driven by the Bristol brass and copper interests. As we have seen, copper ore typically lies at greater depths than the tin deposits, typically well below the water table. If copper mining was to develop in Cornwall the pioneer copper miners would have to find technological solutions to the pressing problem of working below the water table.

From the early 1690s until the 1730s copper mining in Cornwall was dominated by two generations of the Coster family. The members of the Coster family who are of direct interest in this context are John Coster (II) (1647 – 1718) and two of his sons: Thomas (1684 – 1739) and John (III) (1687/8 – 1731).[1] John Coster (II) played a key role the development of copper mining in the South West and has rightly, in the current author's opinion, been described as the "father of Cornish copper mining".[2] John (II) died in 1718, a wealthy and prosperous man, leaving his west of England mining interests to his wife Mary and his sons Thomas and John (III).

John Coster (II) had been involved with Clement and Talbot Clerke's pioneering work on reverberatory copper smelting in the late 1680s, being described as the "chief technical man" of that concern.[3] John Coster (II) was also intimately concerned with the establishment of a very early copper smelter at Upper Redbrook in the Wye Valley, acquiring a sixty year lease for a site in July 1691. There is a suggestion that Upper Redbrook was sourcing a supply of copper ore from the Forest of Dean. However it would appear that the Forest of Dean copper was a very limited deposit; Thomas Cletscher, writing in 1698, noted that it had been quickly exhausted. It is possible that the exhaustion of local copper ores led Coster and the Upper Redbrook smelters to turn their attention to Cornwall.[4] At first they seem to have been content to buy copper ore via agents but it soon became apparent that the quality of ore being dispatched from Cornwall was at best variable and at worst unusable. To source better ore John Coster (II), started buying direct from Cornwall. Initially buying on behalf of Upper Redbrook, he soon became the agent for a consortium of Bristol brass and copper interests usually referred to as the "Bristol Company".[5]

It appears that Costers' main source of supply was "the copper mine near Truro".[6] The mine in question was at Chacewater and was owned by the Boscawen family. Thomas Tonkin, writing in the first third of the eighteenth century, noted that copper mines on the Boscawen Estate were "the first wrought to any purpose in Cornwall"; Pittslooarn, which later became part of Wheal Busy, being first worked in 1681 or 1682, possibly in breach of the Company of Mines Royal's monopoly on copper mining in Cornwall. Metal Work, which lay on the same lode as Pittslooarn, was described by Tonkin as "the mother of copper mines" and, in the years around the turn of the century, yielded Hugh Boscawen*, "more money for copper than most all the mines in the county put together".[7] Yet in spite of the undoubted richness of the Chacewater mines Boscawen was experiencing problems. When Thomas Cletscher, in effect a Swedish industrial spy, visited the mine in 1696 he noted that "the mines are 40 or 50 fathoms deep" and were suffering from water problems.[8] Chacewater would prove to be a notoriously wet mine and a significant challenge for successive generations of engineers; for example in a letter dated August 13th 1778 James Watt comments to his business partner Matthew Boulton: "ye water is reckoned the heaviest in the whole county".[9] Consequently output was suffering and skilled miners were deserting the mine for the copper mine at North Molton. Boscawen had started to drive an adit to drain the workings at depth, but, as Cletscher observed, it was "not yet half carried on".[10] Indeed it would appear that draining the Chacewater mines was too great a challenge for Boscawen as the mine was subsequently abandoned depriving the emergent copper industry of one of its major sources of supply.

In order to secure the Chacewater ore the Bristol Company acquired the lease of the Metal Work and Pittslooarn / Wheal Busy sections of the mine from the Boscawen family at dues of 1-9th,[11] thus becoming directly involved in mining in Cornwall. The Bristol Company would go on to dominate copper mining in the South West during the first third of the eighteenth century. Their interests were largely represented in Devon and Cornwall by the Coster family; John Coster (III) taking up residence in Cornwall by 1714, the better to superintend both his family's and the Bristol Company's mining interests in the region.

To dewater Chacewater mine the Bristol Company initially relied on adits. By the 1690s the use of adits was well known in the South West, having been in use as early as 1309 in the Bere silver mines in Devon.[12] Richard Carew in his *Survey of Cornwall* of 1602 notes that tin works were frequently drained by adits. What makes the Chacewater adits stand out is the scale of the undertaking which dwarfed anything which had gone before. Henric Kalmeter visited the mine on the 4th December 1724 and wrote:

> "To these works twin adits have been driven, the first an old one (Author's note: *presumably Boscawen's part finished adit of 1696*) and the later one, which is no deeper than the first but has cost about £11,000. Because they intended to survey the ground they have sunk shafts in many places and

***During the period under consideration two members of the Boscawen family are of importance: Hugh Boscawen (I) (1625 – 1701) and his son Hugh (II) (1680 – 1734) who was created Viscount Falmouth in 1720.**

continued working in the level at numerous places, by which the work has gone on rapidly. It became dreadfully expensive because of the many shafts and the many workmen who are employed to work on both sides of the shafts and towards each other. Thus they have in less than three years driven an adit over a mile long begun in the lowest ground they could find. What otherwise makes the work expensive is that they have to keep men to pump the water up the shafts instead of driving the level so that the water runs out on its own and without further expense."[13]

The Chacewater adits may be seen as the prototype for the longer, deeper adits which were driven during the eighteenth century to open up copper deposits at depth. To drive adits on this scale required both significant technological and financial advances.

Key to the driving of long adits was the introduction of gunpowder to mining in south west England which significantly speeded up rates of driving and sinking. It is generally accepted that gunpowder was introduced into mining in Cornwall in 1689, possibly at the instigation of the Godolphins. The first documentary evidence for its use is to be found in the Breage parish register:

"Thomas Epsley Senior of Chilchumpton pish (*parish*) of Bath and wills (*Wells*) in Sumersitsheers (*Somersetshire*) he was the man that brought that rare invention of shooting the rocks which came here in June 1689 and he died at the ball (bal) and was buried at breag (*Breage*) the 16 day of December in the year of our lord Christ..... 1689"[14]

Given the Somerset connection it would be reasonable to assume that Epsley acquired his knowledge in the Mendip lead mines. Kalmeter describes the technique of "shooting the rocks" as used in the 1720s:

"The work in the mine is performed either with hammers and wedges, or, when the country rock is very hard, they blast it with gunpowder, boring the hole seventeen or eighteen inches deep or more, and filling above the gunpowder with clay. In some places they find it best to tamp it with the fine gravel which comes out when boring the hole, which is mentioned here because instead of this they recently began the practice of using an iron bar between the plug and the wall of the level so that the plug should not spring out".[15]

Writing at a similar time to Kalmeter, Tonkin observes:

"They had of late had recourse to Gunpowder by boring holes in them, in the nature of mining: and this devise has been likewise attended with many fatal accidents, by powder taking fire too early by a spark from the rod; which hath been of late much remedied by a new method, introduced from abroad, by Major Joseph Sawley as used in mining in sieges, by not making use of the rod at all, but covering the powder and fuse with fine earth, which answers full as well as if it was rammed in."[16]

As to the source of powder used in the South West: Bryan Earl in *Cornish Explosives* notes that the earliest reference to gunpowder being made in Cornwall is 1809; prior to that he suggests that powder was supplied by London or Somerset mills.[17] It comes as no surprise to learn that Bristol and its hinterland was an important centre of gunpowder production during the latter seventeenth and early eighteenth centuries.[18]

The use of gunpowder in laboriously hand drilled shot holes revolutionised Cornish mining in the late seventeenth and early eighteenth centuries.

To make the transition from comparatively shallow tin working to deep copper mining required not only a technological step forward as represented by the introduction of gun powder but also increased financial commitment on behalf of adventurers. As discussed in the introduction a comparable transition had taken place in the Tudor period when the focus moved from alluvial tin to lode tin. In terms of capital, tin streaming was within the realm of the working tinner whilst lode mining was comparatively capital intensive. The capital to develop lode tin mining came from within the South West from wealthy tinners, merchants and local gentry; men like the Boscawens. Whilst the resources of the Boscawen family and their ilk were sufficient to exploit lode tin, or indeed copper, at shallow depths, working copper below the water table and the driving of long adits was a game too rich for local pockets alone. This is where the mercantile strength of Bristol, a city growing ever richer on the back of the recently deregulated slave trade, became important. The huge cost of driving the Chacewater adit, which was way beyond the means of the previous generation of largely local adventurers, demonstrates both the level of investment required to develop a deep copper mine on this scale and the vast financial resources that the Costers and the Bristol Company had at their disposal.

Chacewater was not the only significant adit driving project involving the Costers.

In 1710 a group of adventurers including Francis Basset and John Coster (II) took out a lease to work what became Pool Mine, primarily for copper. Key to the adventure was the driving of a long adit from the Tuckingmill Valley to Pool. Work on the Pool adit had started by January 1711, the adit being driven with gunpowder. Allen Buckley, who has examined the "Penhellick worke" cost books, comments on the rapid progress achieved using drill and blast techniques, with five fathoms a month being typical. Driving the adit was not without incident. For example the cost book records a payment of £5 5s between 1st September 1722 and 19th November of the same year to Doctor Roskruger "for the care of Wm Tresaderne being hurted by shooteing with powder in ye additt end....." That the adit was a success is beyond doubt; by the 1740s Pool Adit was giving the Basset family an annual profit of £11,000.[19]

The ability, both technical and financial, to drive long adits reached its zenith in the "Great County Adit". Inspired by the success of the Pool Adit the County Adit was conceived by William Lemon (1697 – 1764) and John Williams. William Lemon who became the dominant figure in Cornish copper mining during the middle years of the eighteenth century, is believed to have started his career as a clerk to "Mr Coster" (presumably John Coster III).[20] Lemon would play a highly significant role in the adoption of the Newcomen engine in Cornwall. The adit, started in 1748, was initially at least, intended to drain Poldice Mine. This was an ambitious project; to drain Poldice to the western boundary of the sett would require an adit two and a half miles long. By 1756 the adit was one and a half miles long and had reached

The adit portal of the Great County Adit. Work on the adit started in 1748 and eventually extended to nearly forty miles and would drain over sixty mines. ***Photo: Tony Clarke.***

the eastern end of the Poldice sett. In 1767 the "Poldice Adit" reached its goal, the western boundary of the sett. Whilst draining Poldice was the initial aim of the adit it soon began to expand into neighbouring mines, developing into what Buckley describes as a "regional mine drainage system" ultimately serving most of the Gwennap copper district.[21] Richard Thomas in his *Survey of the mining district of Cornwall from Chasewater to Camborne* of 1819 describes the "Great Adit" thus:

> "This adit discharges its waters in the valley near Wheal Friendship and Nangiles, and extends to mines which are as much as four miles in *horizontal* distance from its mouth. Its course upwards from the adit mouth is through the valley by Twelveheads to Hayle Mills. From this line branches go to *Wheal Friendship* and *Wheal Fortune*.
>
> From Hayle Mills go two principal lines: one of them through *Wheal Virgin* to *West Wheal Virgin, Wheal Maid* and *Carharrack*: from West Wheal Virgin to the *United Mines, Wheal Squire*, and *Ting – Tang*. From the United Mines one branch goes to *South Ale and Cakes*, and another to *East Ale and Cakes*.
>
> The other line from Hayle Mills goes north – west through the valley to the eastern part of *Poldice Mine*, where it again divides into two principal lines. One of these goes through *Poldice Mine* westward to *Wheal Quick, Wheal Jewel, Wheal Damsel*, and *Wheal Hope*. From Wheal Jewel a branch goes by the Great Cross – course (e) to *Tolcarne Mine, Roslabby*, and *Wheal Pink*. From *Wheal Quick* a branch goes through *Wheal Gorland*, and is extending towards *Wheal Clinton*. From the middle of Poldice Mine a branch goes northward through *Wheal Unity* to *Wheal Union*. From the east part of Poldice Mine a branch goes to the east part of *Wheal Unity* and into *Creegbraws Mine*.
>
> The other principal line from the eastern part of Poldice Mine, goes northward through *Creegbraws* to *Chasewater Mine*, and from thence westward through *East Wheal Chance, Halbeagle, Wheal Rose, Wheal Hawke, Wheal Messer, Wheel Peevor; Good Success*, and *Wheal Maria*. From East Wheal Chance a branch goes to *Scorrier Mine*. From Halbeagle a branch to *East Downs Mine*. From Wheal Hawke a branch to *Briggan*, and northwards from that mine towards *Wheal Barberry*. From Wheal Hawke a branch to *South Wheal Hawke, Wheal Chance, Treskerby*, and *Wheal Boys*. From Wheel Peevor a branch to *Wheal Prussia, Cardrew Mine* and *Wheal Derrick*.
>
> The length of the adit, with the various branches ...... amounts to about 26,000 fathoms, nearly 30 miles; and the greatest length to which any branch appears to have been extended from the adit mouth is at Cardrew Mine, which is about 4800 fathoms, nearly 5½ miles. The highest ground it has penetrated is at Wheal Hope, where the adit is 70 fathoms deep at Chilcot's Shaft, and is deeper in the branches extended from hence."[22]

Whilst undoubtedly successful, the driving of adits was only a partial solution to the problem of water management and it was inevitable that copper miners would want to chase their lodes below adit level necessitating mechanical pumping. The need to pump was nothing new, the challenge having been addressed, to a degree, by the tinners as they honed the techniques of lode mining, Carew notes:

> "For conveying away the water they pray in aid of sundry devices, as adits, pumps, and wheels driven by a stream and interchangeably filling and emptying two buckets, with many such like, all which, notwithstanding, the springs so encroach in these inventions as in sundry places they are driven to keep men, and somewhere horses also, at work both day and night without ceasing, and in some all this will not serve the turn".[23]

William Pryce, in his seminal work *Mineralogia Cornubiensis* of 1778, echoes Carew's observations:

> "With all the skill and adroitness of our Miners, they cannot go any considerable depth below the Adit, before they must have recourse to some contrivance, for clearing the water from their workings. The hand pump, and the force pump, will do well for small depths, and are necessary in the first sinkings into the Lode, before the Stopes can proceed. Next to these the water is drawn to adit by small water barrels, in a core of six or eight hours, they give over drawing by hand, and erect a Whym, which is a kind of horse engine to draw water or work.........
>
> Another water engine is the Rag and Chain.......... Several of these pumps may be placed parallel upon different Stulls, Sallers or Stages of the mine, and are usually worked by hand like those in our navy. The men work at it naked excepting their loose trowsers, and suffer much in their health and strength from the violence of the labour, which is so great that I have been witness to the loss of many lives by it."[24]

Thomas Savery in his *Miner's Friend* of 1702 has this to say on the subject of rag and chain pumps:

> "I have known in Cornwall, a work of three lifts, of about eighteen feet lift, and carrying a three and a quarter inch bore, that cost forty two shillings per diem, reckoning twenty four hours the day, for labour, besides wear and tear on the engines; each pump having four men working eight hours, at fourteen pence a man, and the men obliged to rest at least one third part of that time."[25]

The inadequacy of existing technology, as highlighted by Carew, Pryce and Savery, is well illustrated by the example of the Marquis Copper Mine on the Devon

Woodcut from Agricola's De Re Metallica of 1556 showing a typical man operated rag and chain pump of a type that would have been common in Cornwall during the eighteenth century. The Costers would turn the concept of the of the rag and chain pump on its head with their patent water engine of 1714.

bank of the River Tamar:

> "She was first discovered seventeen years ago (1707) on the bank of the river, when some workmen got together, took a set or lease of 20 fathoms from the water's edge from the owner of the land, and drove a level from the lowest point. They found some copper ore, and the work or place was called Bedford. Immediately after this some enterprising adventurers formed a company and took a sett of the ground above Bedford. They drove a level, which is vertically seven or eight fathoms above the former one, and more than sixty fathoms long, though it was driven in such a way that it was made parallel with the first

level. Thereupon both workings were united and were and still are called the Marquis. When those holding the second sett or tack note began work they first found tin ore upon the back of the lode, or upmost in the ore ground. But when they sank on it they found copper, and that in much larger quantities and of a better kind than in the first level. The ore shoot runs exactly east and west, and up the hill from the river. Probably as the water took the upper hand, they had sixty men employed to pump it out, but the labour became to great and was abandoned.......”[26]

The Marquis lay idle until 1723 when the lease was acquired by the “Copper Company in Bristol” who promptly installed a “water engine” at the end of the upper level. The engine was an overshot waterwheel; recent archaeological investigation suggests that the wheel was in the region of thirty feet in diameter and three foot breast.[27] The water was laundered in along the upper adit level; whilst tail race water and the water pumped from below deep adit went out via the deep adit to the river. At the time of Kalmeter’s visit on 13th November 1724 a shaft had been sunk to a depth of sixteen fathoms, the intention being to put in pumps once they had cut paying ore ground.[28]

The use of waterwheels to drive pumps was far from unknown in the west of England; waterwheel driven pumps having been in use on the Bere Ferrers mines since 1480.[29] What was innovative was the use of large, overshot wheels; an innovation credited to John Coster (II). In this context William Pryce wrote:

“About four score years back, small wheels of twelve or fifteen feet diameter, were thought the best machinery for draining the Mines; and if one or two were insufficient, more were often applied to the purpose, all worked by the same stream of water. I have heard of seven in one Mine, worked over each other. This power must have been attended with a complication of accidents and delays. However, soon after the above date, Mr John Costar, of Bristol, came into this country, and taught the natives an improvement in this machinery, by demolishing those petit engines, and substituting one large wheel of between thirty and forty feet in their stead.”[30]

Pryce provides a good description of a large, overshot pumping wheel:

“The water wheel with bobs, is yet a more effectual engine; whose power is answerable to the diameter of the wheel and the sweep of the cranks fixed to the extremities of the axis. Over them two large bobs are hung on brass centre gudgeons supported by a strong frame of timber, and rise and fall according to the diameter of the sweep of the cranks, or of the circle they describe. To each crank is fixed a straight half split of balk timber, that communicates with each bob above: at the other hand or nose of the bob over the Shaft, a large iron chain is pendant, fastened to a perpendicular rod of timber that works a piston in an iron or brass cylinder, called the Working Piece: the quantity of water exhausted, will be in proportion to the bore of the working piece, and the number of times which the embolus works up and sown in a given space ...... this kind of engine is the most eligible, where grass water is plenty, and to be had for a small rent."[31]

In addition to the Marquis wheel Kalmeter recorded details of other overshot wheels being erected on mines with a Coster / Bristol Company connection:

"On 3rd December 1724 Kalmeter visited Tolgus Downs copper mine run by Thomas Coster and a Captain Miners. Kalmeter noted that Tolgus Downs was one of the oldest copper mines in Cornwall, having been "worked continuously for copper for nearly thirty years". At the time of Kalmeter's visit an "engine to draw water" was being installed to pump water from the bottom of the mine to adit. The engine itself comprised of a thirty four foot diameter overshot waterwheel, typical of the Coster's thinking and practice. A pair of cranks attached to the axle drove two sets of pumps. In December 1724 the bottom of the engine shaft lay eleven fathoms below adit although the engine had the potential to pump from a depth of twenty four fathoms."[32]

John Coster (III) was responsible for installing "water engines" at the extensive North Molton Copper Mine in North Devon. Kalmeter, who visited the mine of 28th December 1724, described the engines in some detail:

"In both workings water-engines were put in by the aforementioned John Coster, and the mine is worked by the so-called Bristol Company. To put these machines to work they have water flowing from the river a good bit higher upstream. In the western working it consists of a wheel 20 feet in diameter, to the end of whose axle is fixed a crank or double iron cranks, which lifts two sets of rods 4 fathoms long and deep, being cast of iron in such a way that when one rod is lifted up by the one crank the other is forced down by the other crank. Each one of these rods or bars works a so-called bob or arm 4 fathoms underground, which as mentioned goes up and down, and these arms also work two pumps and pump-rods in the excavation, which is 16 fathoms deep. The water which comes out of the pumps goes into two cross- pipes or pump launders, which lead across into a larger pipe running into the level, by which means it is lead of to the stream.

The machine in the eastern work is of the same design and construction suited to its work, except that the water wheel is 24 feet in diameter; the working iron rod also has the rods and bobs, or arms at 20 fathoms depth. But in addition there are two half – wheels on the side of the shaft, which are moved up and down by two iron rods, which are attached at one end to the iron crank, and at the other end to the half wheel; they are thus in constant motion upwards and downwards. The one of these half – wheels works in the same was as mentioned under Chasewater Mine, a pump at 30 feet and another at 40 fathoms, or down to the bottom of the mine."[33]

Where appropriate the large overshot water wheel was an ideal solution to the problem of mine drainage as is demonstrated by their wide scale adoption right up to the twentieth century; however they were by no means a universal solution. Pryce summarises the problem elegantly:

"Happy it would be for the mining interest, if our superficial streams of water were not so small and scanty; but the situation of our Mines, which is generally in hilly grounds, and the short current of our springs from their source to the sea, prevent such an accumulation of water, as might be applied to the purpose of draining the Mines; and of course the value of the water is the more enhanced. There are very few streams, which are sufficient to answer the purpose in summer, as well as in winter, so that many engines cannot be worked from May to October."[34]

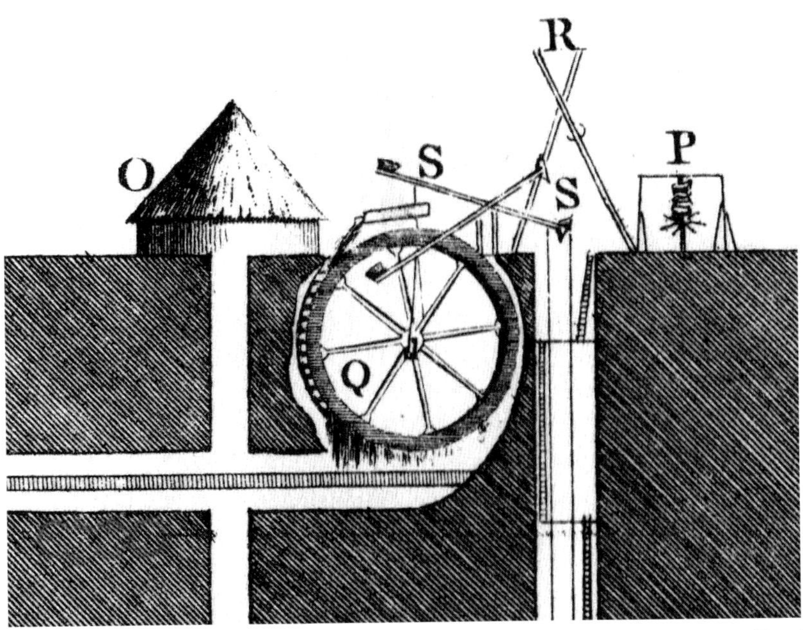

A typical eighteenth century bob engine, this example was at work at Bullen Garden in the 1760s. It comprises a large overshot waterwheel driving two sets of lift pumps via a pair of bobs connected to cranks on the wheel's axle. The bob engine was just one of the Coster family's contributions to mining in the south west.
Image courtesy Tony Clarke.

A consequence was that adventurers were often charged exorbitant sums for the use of a watercourse. In a letter dated 18th January 1757 William Lemon commented:

"I have heard of £50 – a month, or more, paid for a water Course for driving a Wheel Engine."[35]

As Pryce suggests some mines with inadequate water supplies often had to be abandoned periodically. Kalmeter cites the example of "Frenches Work" near Roseliddon:

"Mr Coster now runs the work on tribute........ The water is pumped up by hand from the bottom to the aqua duct (*adit*), and it can be noted that this is one of the workings called a summer work, since it can only be worked in summer, water hindering them in winter."[36]

Whilst abandoning mines on a seasonal basis was sometimes the only option, the situation would arise when a mine, which lacked a sufficient water supply to drive a large overshot wheel, was so rich that abandoning it for months on end was not a viable option. The Costers' response to this problem was to develop an ingenious water engine which required less water than a large overshot wheel. John (II) and his son John (III) patented their water engine in 1714:[37]

"AD 1714 ..... No 397
Coster and Coster
Engine for drawing water out of mines.
Anne, by the Grace of God &c. To all to whom these presents shall come, greeting: Whereas Our trusty and wellbeloved John Coster, of Redbrooke, in Our county of Gloucester, gentleman, and John Coster, Junior, of Redruth in Our county of Cornwall, gentleman, have by their humble petition represented unto Us, that they have with great paines and expenses invented an brought into perfection an engine for drawing water out of deep mines, much cheaper and more effectuall that the usual ways, by water wheeles; in which new invented engine the water that drives it is carried down the shaft or pit to the adit in bored elemes and mettall cillinders, and by that means circulates a chain with bobbs through the said cillinders, which chain, depending on a peculiar sort of wheele, workes a cranke at each end of the axell thereof; which said engine, with the application thereof for the drawing water with sucking plumps, chaine plumps, and forcing plumps, is a new invention not formerly known and practised in England, and being made for lesse charge than water wheeles, and requiring much less room underground, as well as less charge to keep it in repair, and that it will wit the same stream of water draw at least one third part more water in quantity from the same depth, or one third in depth

more than water wheels with the same stream can or will do, the said engine will of great use for such mines as have been left off for want of ability to draw the waters off them, the recovery and working of which will be of great advantage to the nation in generall as well as to the Lords of the Royalties and persons adventuring in such mine; and having therefore humbly prayed us to grant them Our Royal Letters Patent for the sole use of their said invention for fourteen yeares."[38]

The Costers' water engine was, in effect, a rag and chain pump albeit, in an elegant display of lateral thinking, used as the prime mover rather than as a pump. One of these engines, possibly the prototype, may have been erected on the Pittslooarn / Wheal Busy section of Chacewater Mine as early as 1712 – 1714. We are indebted to Henric Kalmeter who recorded comprehensive details of the engine which he examined on 4th December 1724. As described by Kalmeter the Chacewater engine worked on the Costers' rag and chain principle applied on a large scale; the "working column" extended to a depth of eighteen fathoms. Technically speaking the description is confused; it appears that Kalmeter failed wholly to understand the principles of the machine and the description should be read with that in mind:

"If there is anything which should be noted first about this engine, it can be said that the wheel which gives motion to the whole machine is built over the middle of the shaft, though a little below surface. It is six feet in diameter, and over it runs an endless chain or *catena perennis*, down on one side and up on the other. On this chain there are forty balls made of strong yarn and provided with leather almost like a sucker in a pump. These balls and chain work vertically upwards (*Author's note: for the machine to work read "downwards"*) in a pump column, which stands on its end in the shaft and is eighteen fathoms long and fourteen inches in (external) diameter. The lower end of this pump column goes into a brass cylinder, which stands at the bottom and which is seven feet high and eight and a half inches in diameter, through which cylinder and pump column the aforesaid balls, which are the same diameter above as the cylinder, or eight and a half inches, but smaller and round underneath and nearly of a conical form run down and then up again through the cylinder. At the bottom they go through a little trough or gutter, which lies horizontally, and up over the other side of the wheel, the balls coming between the teeth of the wheel, which are either of iron or are strongly reinforced with iron, which turned continuously. Close in to this first mentioned wheel is another one, also cast in iron and of three feet in diameter, which with its cogs engages with the teeth on the first wheel. This smaller wheel does not serve to give any motion to the machine, but only to hold the first named wheel and the whole workings in an even, steady slow motion. Further from this wheel, though on the same axle as the other wheel and near the wall of the shaft runs a third wheel, which is thirteen feet in diameter and has 1,100 pounds of lead on its rim, serving only to balance or counter the weight of the whole machine.

But to return to that which gives movement to the aforementioned wheel and machine. It can be seen that it is done by falling water, which is lead through a level underground and into a pump column which stands on its end, right beside the first mentioned one. On the side of this second pump column, about five feet down, is an opening or hole through which the water runs into the first one. It is then drawn up by the aforementioned balls, which are partly assisted by the weight of eighteen fathoms of descending chain, pressing with the weight and steady fall of the one after the other, there being, as noted a fall of eighteen fathoms (*Author's note: Kalmeter has misunderstood the operating principle: it is the weight of the water in the pump column acting of the balls which creates rotary motion and not the weight of the chain and balls lifting the water*). The balls are arranged at such a distance from each other so that one does not come up and out of the brass cylinder at the bottom before the next one has come into it. Furthermore, when it is necessary to stop the machine, on the second pump column where the water runs in there is a slide which can block the pipe completely. By a rope, which is fixed to a small iron hook, the slide can be raised over the opening to stop the water flowing onto the balls in the first pump column. Then it flows down, running off into a deeper adit without the help of the machine. Alternatively, the slide can be let down below the opening, when the water flows into the other column and on to the balls when the machine is set working again. By means of this slide one can also regulate the sped of the machine, by raising it more or less over the opening in the column, thus letting in more or less water. To show either the rise or fall in the depth of water down in the mine, shaft or sump, they generally have a buoy or float which rises and falls with the water. A rope goes from it up through the shaft to a lever, on the other end of which is a counter weight or lead, which accordingly sinks of rises and serves on its own to regulate the slide on the pump column, being fixed to an arm by which the slide and its iron weight and rope can be let up or down.

This is the action and movement of the machine. The machine is arranged so that on each side of the first mentioned wheel of six feet in diameter and on each end of its axle is a strong iron crank, which, as the axle revolves, drives the rod up and down in the pump, which pumps up the water. Its stroke is five or six feet. On one side of this wheel are the columns for the rag and chain pump, and on the other, beyond the shaft, an excavation in the country rock for the so called half wheels. (*Author's note: the term "half wheel" is somewhat misleading; a better term would be "arch head" as in the end of the beam in atmospheric and Boulton & Watt engines*). These are only built in a half part of a circle eight feet in diameter, which are fitted on iron rods with the half wheel at one end and the cranks on the other. They move up and down. From the

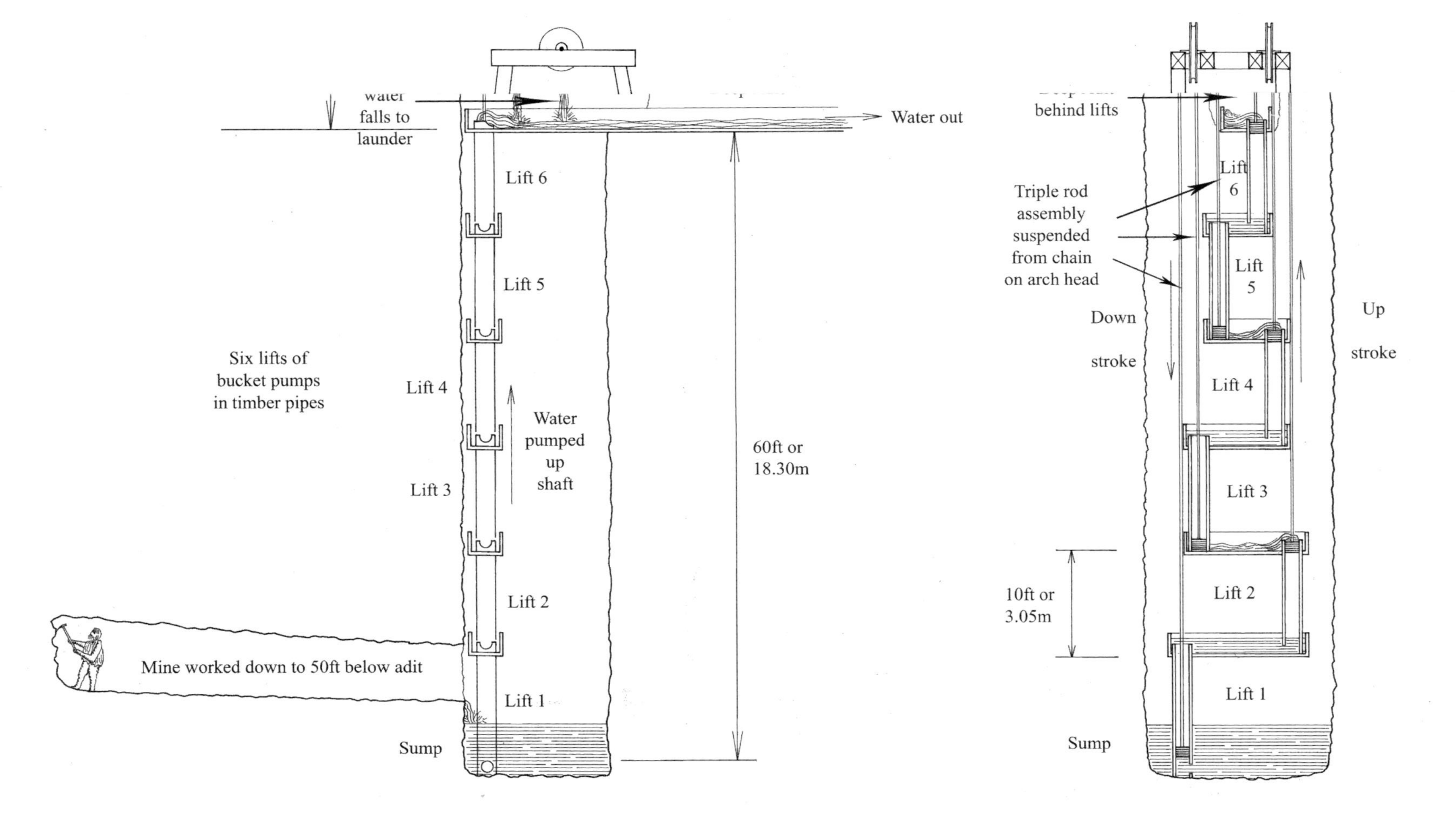

Conceptual reconstruction drawing of the Costers' patent water engine. The drawing does not represent a specific engine. The design is an interpretation of the engine by the author and Robert Waterhouse based on Kalmeter's description of the Chacewater engine and Desaguliers illustration. Supporting timber work has been omitted to avoid confusion, as has the regulating mechanism. ***Drawn by Robert Waterhouse.***

and a half fathoms."[43]

Relistian Mine on Gwinear Downs was visited by Kalmeter on 1st December, 1724 at which time it was being worked to a depth of twenty fathoms for both tin and copper. At its peak Relistian had been a major tin producer, Kalmeter noted that it produced 200,000 pounds of tin metal in 1717 and that the lode had been forty feet wide in places. At its greatest extent Relistian had reached a depth of seventy five fathoms and had been drained by a water engine of "the same kind as the one at Chasewater mine". Unfortunately the lode at depth was found to be weak and narrow and by 1720 the water engine had been removed.[44]

Whilst the Costers' patent water engine did, to a limited extent, address the problem of the poor supply of river and stream water in Cornwall the real solution to the problem lay in the application of steam to mine pumping.

Chapter 1 References

1. www.bittonfamilies.com, Day J, 1977
2. Barton D. B., 1961; Day J., 1977
3. Day J., 1977
4. Jenkins R., 1942
5. Brooke J. 2001; Morton J., 1985
6. Cletscher T., 1696
7. Morton J., 1985, Tonkin MSS
8. Cletscher T., 1696
9. Cited in Dickinson H. W. & Jenkins R., 1927
10. Cletscher T., 1696
11. Brooke J, 2001
12. Claughton P, 1994
13. Brooke J. 2001
14. Cited Earl B., 1978
15. Brooke J., 2001
16. Tonkin MSS B
17. Earl B., 1978
18. Buchanan B. J., 2000
19. Buckley A., 2016
20. Polwhele R., 1831
21. Buckley A., 2016
22. Thomas R., 1819
23. Carew 1602
24. Pryce W., 1778
25. Savery T., 1702
26. Brooke J., 2001
27. Waterhouse R. E., *pers comm*
28. Brooke J. 2001
29. Claughton P., 1994

30. Pryce W., 1778
31. *Ibid*
32. Brooke. J. 2001
33. *Ibid*
34. Pryce W. 1778
35. Borlase letter books
36. Brooke J., 2001
37. Stewart R. J., 2015, Woodcroft, B.,1862
38. Bennet Woodcroft, 1862
39. Brooke J., 2001
40. *Ibid*
41. Tonkin MSS H
42. Brooke J. 2001
43. *Ibid*
44. *Ibid*

Chapter 2
Thomas Savery, the "miner's friend" and the patent "for raising water by the impellent force of fire"

The first inventor to turn his attention to the application of steam pumping to Cornish mining appears to have been "Captain" Thomas Savery (1650 (?) – 1715). Savery's biographical details are somewhat sketchy. He was born into a family of successful Devon merchants in the middle years of the seventeenth century It is uncertain where he acquired the title "Captain"; whether military, nautical or mining honorific is moot, certainly nineteenth century sources appear to be at odds. *The Gentleman's Magazine* of July – December 1839 notes that "he was by profession a military engineer". He also had an interest in nautical matters: In 1686 he took out a patent for rowing vessels with paddle wheels and in 1698 published *Navigation Improved.* In 1705 Savery was appointed Treasurer to the Admiralty Commission for the Sick and the Wounded, an important post with a salary of £200 per annum which he held until 1713, two years before his death.[1] In contrast Farey writing in the 1820s observes that:

> "He is said to have been called Captain by the miners in Cornwall, in consequence of his being employed to drain the water for them; it is still their custom to give the title of Captain to engineers."[2]

Savery's booklet *The Miner's Friend* of 1702 demonstrates that he had knowledge of Cornish mining practice presumably gained first hand.

Like most things pertaining to Savery the origins of his engine are somewhat obscure. Desaguliers, in the second volume of his *Experimental Philosophy* of 1744, gives an entertaining, albeit jaundiced, view:

> "Captain Savery, having read the Marquis of Worcester's Book, was the first who put in practice the raising of Water by Fire, which he proposed for the draining of Mines. His Engine is describ'd in Harris's Lexicon (see the word Engine) which being compared with the Marquis of Worcester's Description, will easily appear to have been taken from him; tho Captain Savery denied it, and the better to conceal the matter, bought up all the Marquis of Worcester's Books that he could purchase in Pater – Noster Row, and elsewhere, and burn'd 'em in the presence of the Gentleman his friend, who told me this. He said he found out the Power of Steam by chance, and invented the following Story to persuade People to believe it viz. that having drunk a Flask of Florence at

a Tavern, and thrown the empty Flask upon the fire, he call'd for a Bason of Water to wash his Hands, and perceiving that the little Wine left in the Flask had filled up the Flask with Steam, he took the Flask by the Neck and plunged the Mouth of it under the Surface of the Water in the Bason, and the Water of the Bason was immediately driven up into the Flask by the Pressure of the Air. Now he never made such an Experiment then, nor designedly afterwards, which I thus prove:........"[3]

Whatever the truth behind the development of his engine, on 25th July 1698 Savery was granted a wide reaching patent (No. 356) "for Raising water by the impellent force of fire". The patent was initially granted for the "usual" fourteen years from 1698, however an Act of Parliament of 1699 entitled *"An Act for the Encouragement of a new invention of Thomas Savery for raising Water, and occasioning Motion to all Sorts of Mill – work, by the Impellent Force of Fire"* (usually referred to as The Fire Engine Act) extended this period for a further twenty one years, the patent expiring in 1733.[4] The extension of the patent to 1733 was to have a significant impact on future development of the steam pumping engine.

In 1699 Savery demonstrated a model of his engine to the Royal Society; although it was not until 1702 that he was able to inform the Royal Society that he had completed a full size engine:

"To the Royal Society. At the request of some of your members, at the weekly meeting, at Gresham – college, June the 14th, 1699, I had the honour to work a small model of my engine before you, and you were pleased to approve of it. Since when I have met with great difficulties and expense, to instruct handicraft artificers to form my engine according to my designs; but my workmen after so much experience, are become such masters of the thing, that they oblige themselves to deliver what engines they make me exactly tight and fit for service, and as such I dare warrant them to any body that has occasion for them."[5]

The Post Man of 19 – 21st March 1702 contains the following notice:

"Captain Savery's Engines which raise Water by the force of Fire in any reasonable quantities and to any height being now brought to perfection are ready for publick use. These are to give notice to all Proprietors of Mines and Collieries which are incumbred with Water, that they may be furnished with Engines to drain the same, at his Workhouse in Salisbury Court, London, against the Old Playhouse, where it may be seen working on Wednesdays and Saturdays in every week from 3 to 6 in the afternoon, where they may be satisfied of the performance thereof, with less expense than any other force of Horse or Hands, and less subject to repair."[6]

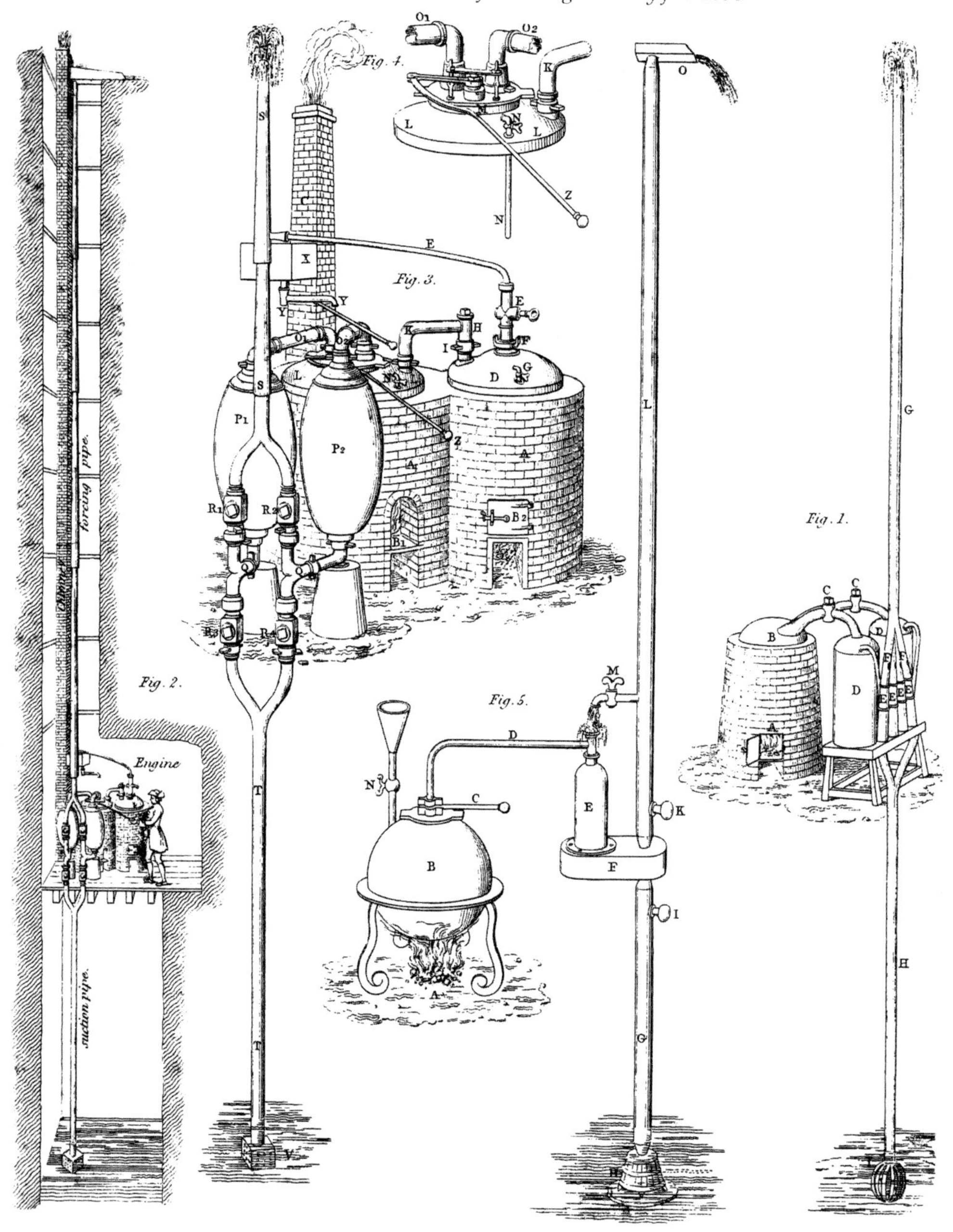

Varieties of Savery engines from John Farey's *Treatise on the Steam Engine* of 1827.
Image courtesy of Tony Clarke.

Savery's engine relied on both atmospheric and steam pressure for its operation. In principle the engine was remarkably simple having only three moving parts: the steam inlet valve or "regulator" and two non return or "clack" valves. At the heart of the engine was a steam /water "vessel" or "receiver". The receiver had connections to the boiler and the water inlet or "sucking pipe" and the water outlet or "forcing pipe". In operation the steam valve was opened allowing steam to pass into the receiver. The valve was then closed and the steam in the receiver condensed creating a partial vacuum. Because the pressure in the receiver was lower than atmospheric pressure water from the sump or cistern would be raised up the sucking pipe, past the inlet clack, filling the receiver. The theoretical maximum of such a lift would be 33.9 feet, although in practice this would be typically between twenty and thirty feet. The operator was able to tell when the receiver was full of water by touch; a full vessel would be cold, an empty vessel hot. Once the vessel was full of water the steam valve would be opened and steam at a higher pressure than atmospheric pressure would enter the receiver causing the water to pass through the outlet clack and up the outlet or forcing pipe. The cycle would then start again. The height that the water could be forced upwards would depend on the pressure of the steam, which in turn was dependent on the quality and strength of both the boiler and receiver. Savery envisaged that his engine would be able to lift water from between fifty to eighty feet. If greater depths were to be reached, for example in mine shafts, multiple engines would be used shammeling from one to another.[7]

Savery saw a number of applications for his engine including pumping water for mills and gentlemen's houses, civic water supply, draining fens and marshes. He also envisaged that his engine "may be made very useful to ships, but I dare not meddle with that matter, and leave it to the judgement of those who are the best judges of maritime affairs".[8] Savery was very keen for his invention to be adopted for mine pumping. To this end in 1702 he published a pamphlet entitled "*The Miners Friend or an engine to raise water by fire described*" in which he notes:

> "For draining of mines and coal – pits the use of the engine will sufficiently recommend itself in raising water so easy and cheap; and I do not doubt, but that, in a few years, it will be a means of making our mining trade, which is no small part of the wealth of the kingdom, double, if not treble to what it now is. And if such vast quantities of lead, tin and coals, are now yearly exported under the difficulties of such an immense charge and pains as the miners, &c. are now at to discharge their water, how much more may be hereafter exported, when the charge will be very much lessened by the use of this engine every way fitted for the use of mines?"[9]

The above reference to tin clearly demonstrates that Savery had at least one eye on the potentially lucrative Cornish market. However in spite of his enthusiastic advocacy of his engine it appears to have made no real impact in Cornwall. As yet no contemporary evidence has emerged that a Savery engine was erected on a Cornish mine. Pryce mentions Savery in passing but makes no comment as to the application of his engine in Cornwall.[10] There are several nineteenth century references to Savery engines in Cornwall; all of which carry the whiff of folklore. The earliest, rather tangential, reference found by the current author is in a paper presented to the Royal Cornwall Geological Society in 1824 by Joseph Carne:

> "The first steam – engine in Cornwall was erected at Huel Vor, a tin mine in Breage, which was at work from 1710 to 1714. Whether this engine was Savery's or Newcomen's is doubtful, as Newcomen's engine does not appear to have been much known before 1712:"[11]

Writing in 1854 James Hamilton informs us that "Savery's first engine was erected at Creegbraws", unfortunately without quoting any source.[12] *One and all*, an obscure and short lived Cornish journal published in the late 1860s contains an interesting, albeit unreferenced, snippet viz: " .... about 1700, Mr Thomas Savery had erected a steam engine...... at Wheal Vor".[13]

Barton, in *The Cornish Beam Engine,* suggests that a Savery engine might have been trialled at Wheal Vor although he fails to quote his source. He, probably rightly, concludes that "it is doubtful whether a trial at such a mine as Wheal Vor, deep even in 1705 or thereabouts, would have served much useful purpose".[14] Pole in his *Treatise on the Cornish pumping engine* of 1844 comments:

> ".....All his (*Savery's*) arguments failed to induce Cornish miners to avail themselves to his offers, or at least no record or tradition of any such use of the engine has been preserved ..... The objections to its use in deep shafts were very formidable, and such as the miner's, who Savery tells us were by no means unaccustomed to weigh the merits and defects of "new inventions of this nature" foresaw would be fatal."[15]

Savery's failure to induce Cornish adventurers to adopt his engine resulted from a number of inherent weaknesses in his design; these are neatly summarised by Pole in his *Treatise*:

> "The principle of these objections may be enumerated.
>
> 1st. To supply Savery's engine, it was necessary to divide a deep shaft into a series of lifts of about 15 fathoms each, this being the greatest height one engine could be made to raise water with safety. A separate engine, with its boiler and appurtenances complete was then required to be fixed at each lift, in an excavation made at the side of the shaft for the purpose ..... ; this involving much inconvenience from the situation of the engines, and the great outlay of

money from the number required.

2nd. Another consequence of this arrangement was, that as each engine formed an essential part of the whole series, if one was deranged or ceased working from any cause, the whole process of draining was stopped, the works were flooded, and some of the other engines probably drowned; the accident itself thus effectually preventing the application of the means best calculated for its remedy.

3rd. To lift water 15 fathoms, or 90 feet, would have required steam of a pressure of upwards of 30 lbs. on the square inch above atmosphere (supposing 20 or 30 feet of the lift to be performed by condensation), involving a serious risk and danger to the works and the men engaged in them from the possibility of the boilers bursting; particularly as in those days the vessels were not so well made as present, and as Savery does not seem to have been acquainted with the use of the safety valve .....

4th. The expenditure of coals with Savery's engine was very considerable; so that no adequate advantage of economy was offered to compensate for the inconveniences of its use, and the outlay of capital required for its establishment. This objection had particular force in Cornwall, where coals were dear .....”[16]

Where we do have evidence of the engine being tried in a mining context it proved woefully inadequate: In about 1706 Savery erected one of his engines at a place called Broadwaters near Wednesbury in Staffordshire to drain a large pond which was seeping into neighbouring coal pits, however

“the engine.....could not be brought to perfection, as the old pond of water was very great, and the springs many and copious, that kept up the body of it; and the steam when very strong tore the engine all to pieces. After spending much time and money, Savery was obliged to give up the undertaking, and the engine was laid by as useless.”[17]

As might be expected for such a pioneering invention Savery's engine was a curious device, arguably better suited to domestic rather than industrial applications. Desaguliers may well be quoted in conclusion:

“Captain Savery made a great many Experiments to bring this Machine to perfection, and did erect several, which rais'd Water very well for Gentlemen's Seats; but could not succeed for Mines, or supplying Towns, where the Water was to be raised very high and in great Quantities: for then the Steam requir'd being boil'd up to such a Strength, as to be ready to tear all the Vessels to pieces.”[18]

Whilst his engine singularly failed to impress Cornish adventurers, Savery would have an ongoing influence on the subsequent history of the atmospheric engine via his patent.

Chapter 2 References

1. Allen J. S. & Rolt L. T. C., 1997; Farey J., 1827; Anon, 1839; Savery T., 1702
2. Farey J., 1827
3. Desaguliers J. T., 1744
4. Allen J. S. & Rolt L. T. C., 1997; Farey J., 1827; *House of Commons Journal* Vol. 12, 1699; Pole W., 1844
5. Savery T., 1702
6. Cited in Allen J. S. & Rolt L. T. C., 1997
7. Savery T., 1702
8. *Ibid*
9. *Ibid*
10. Pryce W., 1778
11. Carne J., 1828
12. Hamilton J., 1854
13. Anon, 1868
14. Barton D. B., 1966
15. Pole W., 1844
16. *Ibid*
17. Anon 1871; Bagnall J. N., 1854
18. Desaguliers J. T., 1744

Chapter 3
Thomas Newcomen and the atmospheric engine

If Savery failed to make any significant impact on mining, quite the reverse could be said of the next engineer who turned his mind to the application of steam power to mine drainage, the "Dartmouth blacksmith" Thomas Newcomen (1663 – 1729). Like Savery, Newcomen was a Devonian; however, unlike Savery, Newcomen successfully developed a workable engine. Newcomen's invention of the atmospheric engine is one of the most important events in the history of technology.The successful practical application of a reciprocating piston in a cylinder is Newcomen's great legacy, underpinning the industrial revolution. Thomas Newcomen was truly one of the architects of the modern world.

Newcomen, a lifelong Baptist, carried on business as an ironmonger in Dartmouth, supplying items from a simple nail up to large quantities of iron, for example in 1698 / 9 Newcomen purchased twenty five tons of iron from the Foley ironmasters in Worcestershire. Newcomen also supplied tools to Cornish mines, where he would no doubt have become aware of the challenge of keeping deep mines drained. In 1734 Marten Triewald, who knew Newcomen and had erected Newcomen engines, wrote (in Swedish):

> "Now it happened that a man from Dartmouth, named Thomas Newcomen, who had no knowledge whatsoever of the ideas of Captain Savery, had at the same time made up his mind, in conjunction with his assistant, a plumber by the name of Calley, to invent a fire - engine for pumping water from the mines. He was induced to undertake this by considering the heavy costs of drawing water by means of horses, which he found in use at the English tin mines. Mr Newcomen often visited these mines in the capacity of a dealer in iron tools which he used to supply to many of the tin mines."[1]

With regard to engine development Triewald notes that "the work on the prototype of the fire – engine was carried on for ten years altogether".[2] If one accepts, and this is a moot point, that the first engine built by Newcomen was the Dudley Castle engine of 1712, this means that Newcomen must have commenced his experiments in around 1702.

Interestingly Newcomen never patented his invention preferring instead to work

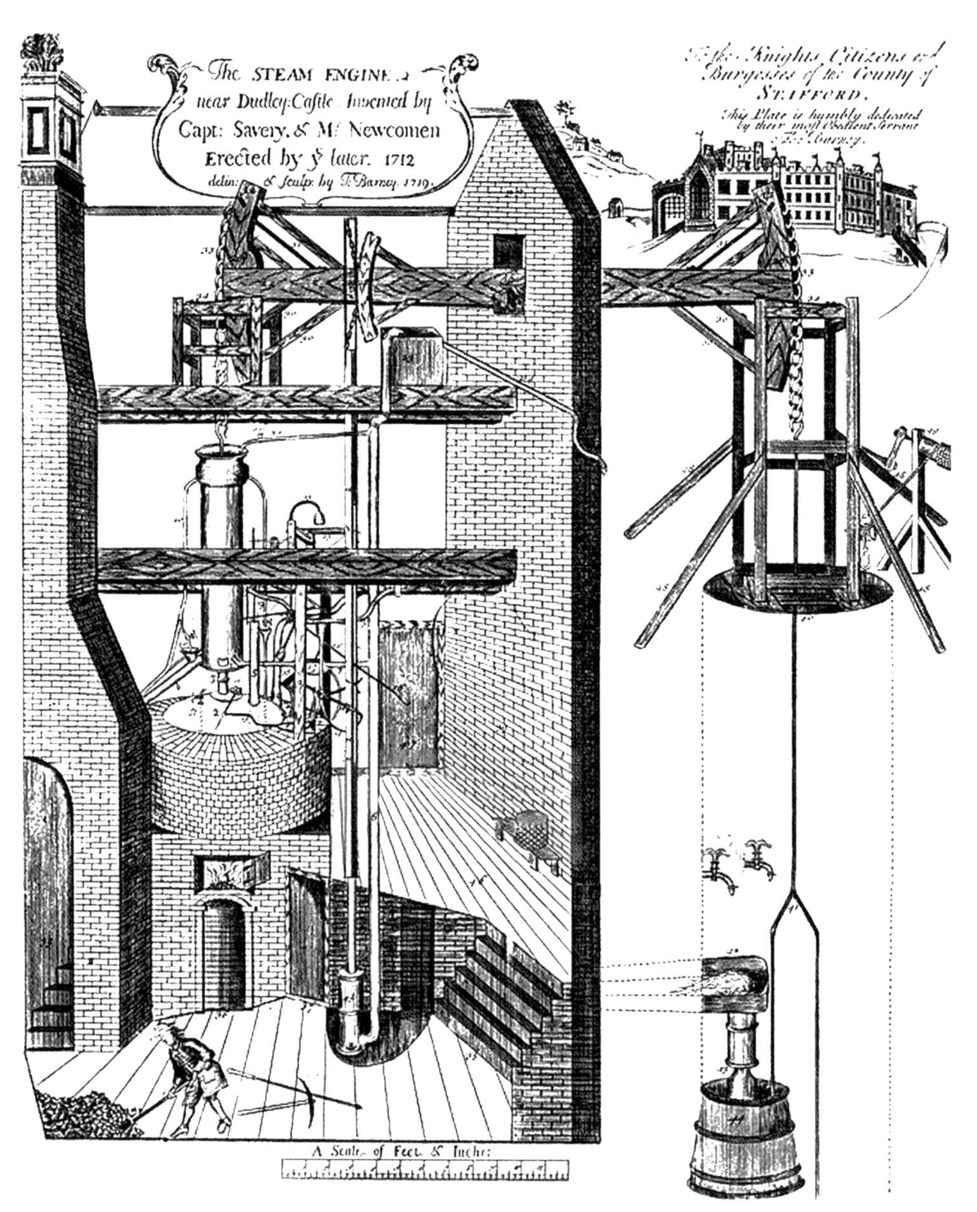

Barney's 1719 illustration of the Dudley Castle engine of 1712. This engine may be considered as the prototype for engines erected in Cornwall.

within Savery's original patent. The reason for this is, at this distance in time unclear. In the 1870s the Patent Office expressed the opinion that Savery's patent was not robust with regard to his claims to Newcomen's engine:

> On comparing Savery's with Newcomen's engine no similarity in construction or operation can be discovered between them; except the regulator plate, and use of steam to actuate both. It would seem, therefore, hardly credible, that Savery

should have attempted to claim, under his patent of 1699, all the profits of Newcomen's engine, as if it had been his own invention; and more particularly, while the Marquis of Worcester's patent was in existence, that might have been bought out to demolish his own. Yet it is certain, notwithstanding, that Savery and his representatives successfully enforced this extraordinary claim, and for thirty – five years imposed the most grinding terms on their lessees, without meeting any resistance."[3]

Whilst Newcomen might have been able to successfully challenge and overturn Savery's patent it may be that Newcomen, taking a pragmatic view, felt that he would be better served working under the protection of Savery's rather wide reaching patent rather than incurring the cost of securing his own. This argument becomes increasingly attractive when one recalls that Savery had managed to secure a twenty one year extension on his patent. The extension of the patent to 1733 may well have come down to the fact that Savery, being a "gentleman" and known in the corridors of power, could apply a degree of influence that Newcomen, an artisan, could not.

On Savery's death in 1715 the rights to the Savery patent passed via his wife Martha (who lived to the extraordinary age of 104) to one John Meres. Meres set up a joint stock company to manage the patent, the company being known as the "The Proprietors of the Invention for raising water by fire". Newcomen appears to be one of the original "Proprietors", although in latter years his interests appear to have been looked after by Edward Wallin.[4]

Anyone wishing to erect a Newcomen engine had to pay a royalty to the "Proprietors". Typical of such royalty agreements was the one entered into between the "Proprietors" and the Laird Wauchope in 1725 who wished to erect a Newcomen engine with a twenty eight inch cylinder and a nine foot stroke. In return for permission to erect the engine Laird Wauchope agreed to pay £80 per year in quarterly payments for the remaining eight years that the Savery patent had left to run. If the sum remained unpaid for forty days the "Proprietors" reserved the right to:

"..... enter, by their servants, horses, carts, carriages, upon the engine, barrels, boilers, pipes, materials, and other things belonging, sell them for the best price that can be gotten, to pay themselves, and return overplus to Mr. Wauchope."[5]

The technicalities of the Newcomen engine have been discussed at great length and with great facility elsewhere and the current author would advise readers wishing to examine the subject in greater detail to acquaint themselves, in the first instance, with Allen and Rolt's definitive work on the subject. That said, it may not be amiss to provide a brief description of the working of Newcomen's engine. The heart of the engine was an open topped cylinder cast from either brass or, latterly, iron. The cylinder was filled with low pressure steam from a boiler situated below the cylinder, the piston being at the top of the cylinder. Once the cylinder was full of steam the valve between the boiler and the cylinder was shut off. A second valve was opened emitting cold water from a cistern at the top of the engine house in the cylinder. The cold water caused the steam to condense creating a vacuum in the cylinder. The pressure differential between the vacuum within the cylinder and atmospheric pressure acting on top of the piston would cause the piston to descend to the bottom of the cylinder. The piston was attached, typically by chains to one end of a bob pivoted on the wall of the engine house. The up stroke of the piston was effected by the weight of the pump rods in the adjacent shaft.

The earliest published technical description of the operation of a Newcomen engine in a Cornish context may be found in William Borlase's *Natural History* of 1758:

> "The most powerful as well as constant engine hitherto invented is the fire – engine. This engine is now well known to the learned, but as their books do not reach everywhere, and this machine is especially serviceable for working of deep mines, and of great advantage to the publick revenue, a general explication of its principle parts, its powers, and profit to the government, may not be improper. The principle members of this engine are exhibited in the plate annexed ..... viz the cistern or boiler T, the cylinder P, and the bob O, I, turning on an axis which rests in the middle of the wall Y. The following is the process of it's several operations: The cistern T, full of boiling water, supplies steam (by means of an upright tube and valve which shuts and opens) to fill the hollow cylinder P, and expel the air through a horizontal tube S, placed at its bottom (*of*) the cylinder, as the steam rises, and the weight of the mine – water depending from I, K, L, perponderates, begins to fill with vapour and the piston which plays up and down in the cylinder rises, and when it got near the top opens a clack by which cold water is injected and condenses the vapour into nearly the twelve thousandth space which it before occupied, and the cylinder being then nearly empty, the piston of iron edged with tow and covered with water (to prevent any air from above getting into the cylinder) is driven down by the pressure of the atmosphere (with the force of about 17 pounds on every square superficial inch) nearly to the bottom of the cylinder; at this instant it opens the valve which lets in the steam from the boiler T, and then the piston ascends until it opens the condensing clack above, which brings it down again to open the under clack and admit the steam, and thus continues ascending and descending as long as the managers think proper, this process is quick, or otherwise, as the steam is by increase or subtraction of fire made more or less violent, to drive the engine faster or slower. To this piston the end of the bob is fastened by an iron chain, and as the piston descends in cylinder P, this end of the bob is drawn downwards, and *vice versa*; as the end O is drawn down, the other end of the bob I, ascends, and by a chain IK, draws up with it, from an iron or brass cylindrical tube, called a pit barrel through a

tyre of wooden pumps ..... a column of water equal in diameter to the bore of that tube, and in height to each stroke or motion of the piston in the cylinder P, and the sweep of the bob IK."[6]

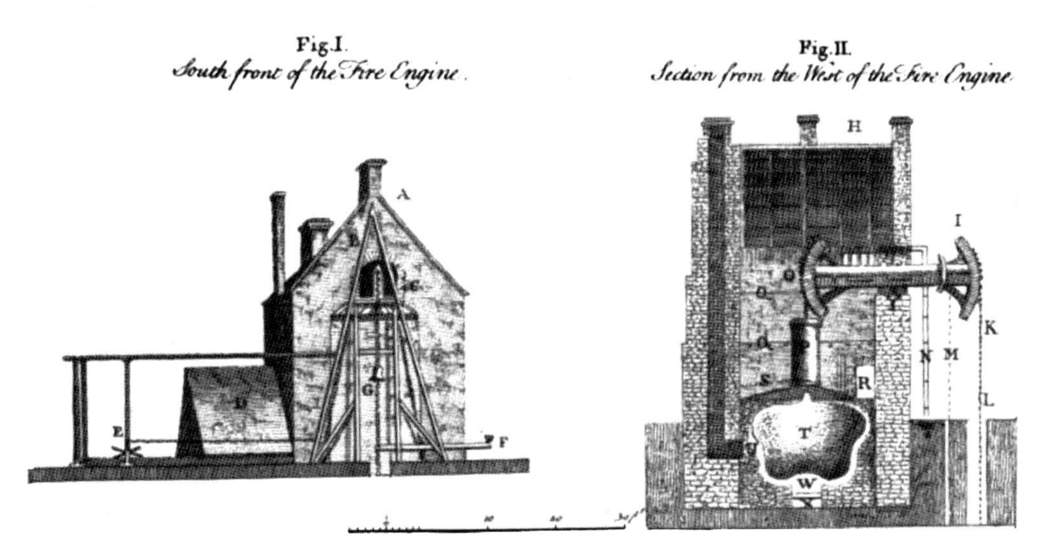

Plate from Borlase's Natural History of Cornwall of 1758 showing the Newcomen engine erected at Pool Mine. ***Image courtesy Tony Clarke.***

The early history of Thomas Newcomen's endeavours are, to a degree, shrouded in mystery. The site of his first engine is a subject for ongoing conjecture and is discussed in the following chapter. It is well established that Thomas Newcomen erected an engine near Dudley Castle on the South Staffordshire coal field in 1712. The 1712 engine represents Newcomen's engine in a well developed form and may be taken as the prototype for later engines of this type erected in Cornwall.

Chapter 3 References

1. Cited in Allen J. S. & Rolt L. T. C., 1997
2. Cited in Allen J. S. & Rolt L. T. C., 1997
3. Anon, 1871
4. Allen J. S. & Rolt L. T. C., 1997
5. Anon, 1871
6. Borlase W 1758

Chapter 4
The First Newcomen engine in Cornwall?

It is as well to start this chapter with an established fact. The first conclusively documented Newcomen engine in England was erected in the vicinity of Dudley Castle in Staffordshire in 1712. This engine is illustrated in Chapter 3 of the current volume. From this point on we will venture into the murky realms of myth, legend, folklore and downright speculation. Marten Triewald wrote the following in 1734:

> "..... Mr Newcomen built the first fire engine in England in the year 1712, which erection took place at Dudley Castle in Staffordshire."[1]

Triewald's statement sounds fairly unequivocal, however Allen and Rolt in *The steam engine of Thomas Newcomen* of 1997 raise a salient point:

> "The design of Newcomen's engine of 1712 was so mature that it is incredible to assume, merely because of the lack of evidence, that there was no intermediate stage of development between it and the small workshop model described by Marten Triewald. Stories are told by Stuart and others of early Newcomen engines with hand – operated steam and injection valves and if there is any truth in these at all it is almost certain that they apply to an engine or engines built before 1712.
>
> For geographical reasons we should expect to find Newcomen making the first full – scale trial in the mining districts of Devon or Cornwall where he was well known."[2]

Circumstantial support for this argument comes from a deposition made in 1748 by one Daniel Hawthorne of Walsall. Hawthorne, aged 60 at the time, had been an erector of Newcomen engines for thirty years. In his deposition he notes that the Old Engine at Tipton, that is to say the 1712 Dudley Castle engine, was "the third built in England......"[3] How much weight one places on Hawthorne's deposition is open to debate, however, in contrast to Triewald, it does admit the possibility of the existence of engines prior to the 1712 Dudley Castle engine.

Allen and Rolt quote A. K. H. Jenkin regarding "a vague tradition" that the first Newcomen engine was erected at Balcoath Mine in the parish of Wendron. They note that the building of this engine was unconfirmed but suggest a tentative date of 1710 – 11.[4] Hamilton Jenkin, in his *Cornish Miner* of 1927, comments:

> "Concerning the first introduction of steam into the county speculation has long been rife. Savery's steam – pump of 1696 had, in 1705, been superseded by one of Newcomen's design. There is a vague tradition that the first of these new engines was worked at "Balcoath," near Porkellis, in Wendron, the steam being raised by turf fuel. This engine, it is said, was subsequently moved to Tregonebris, and then to Trevenen Mine, both ancient tin - works in the Wendron district."[5]

Hamilton Jenkin in turn cites the *Cunnack Manuscript* as the source of this "vague tradition". The *Cunnack Manuscript* was the work of Richard John Cunnack (1826 – 1908) and consists of notes made between 1845 and 1907. Cunnack had moved in Cornish mining circles throughout his long life and was considered an expert on ancient mining in Cornwall.[6] With regard to Balcoath, Cunnack wrote:

> "Balcoath. This is an old mine in the downs north of Porkellis. Tradition reports that the first steam - engine was put up here and that the fuel used under the boiler was turf dug up in the moors adjacent. At a subsequent working another engine was erected here, which was after taken to Tregonebris and after that to Trevenen. This is reported to have been of short stroke, five to six feet, and of small diameter cylinder."[7]

As can be seen Hamilton Jenkin has misinterpreted the *Cunnack Manuscript* confusing the subsequent engine with the original turf burner. Cunnack appears to be the earliest written source for the Balcoath turf burning engine as a candidate for the first Newcomen engine erected in Cornwall (or anywhere else). Given that Cunnack was writing a century and a half after the event he was describing, one has to treat the information with a healthy dose of scepticism.

There is another candidate for a "pre Dudley Castle" engine in Cornwall. In a paper read to the Royal Geological Society of Cornwall in October 1824 Joseph Carne states that:

> "The first steam – engine in Cornwall was erected on Huel Vor, a tin mine in Breage, which was at work from 1710 – 1714. Whether this was Savery's or Newcomen's is doubtful, as Newcomen's engine does not appear to have been much known before 1712......"[8]

This statement seems to be the earliest reference to an engine at Wheal Vor at this period. Unlike the Balcoath engine there is some near contemporary confirmation of an early Newcomen engine at Wheal Vor. On 25th November 1724 Henric Kalmeter visited Wheal Vor and he noted that:

> "as this work was one of the deepest in the county and therefore very wet, a fire engine was erected about seven years ago to draw up the water. It worked for about four years, but as this type of machine, particularly at that depth, is a mass of difficulties, and subject to repairs, it is too costly. The engine was therefore removed, and the mine lies largely idle."[9]

Kalmeter's "about seven years" takes us back to "about" 1717, not 1710 as suggested by Carne. The earliest contemporary reference to a Newcomen engine in Cornwall is an announcement placed by the "Proprietors" in the *London Gazette* of 11th August 1716:

> "Whereas the Invention for raising Water by the impellant force of Fire, authorized by Parliament, is lately brought to the greatest Perfection; and all sorts of Mines &c. may be thereby drained, and Water raised to any Height with more Ease and less Charge than by the other Methods hitherto used, as is sufficiently demonstrated by diverse Engines of this Invention now at Work in the several Counties of Stafford, Warwick, Cornwall and Flint. These are therefore give Notice, that if any person shall be desirous to treat with the Proprietors for such Engines, Attendance will be given for that Purpose every Wednesday at the Sword – Blade Coffee – House in Birchin – lane; London, from 3 to 5 of the Clock; and if any Letters be directed thither to be left for Mr. Elliot; the Parties shall receive all fitting satisfaction and Dispatch."[10]

The phrase "..... engines now at work in the several Counties of .... Cornwall ....." confirms that an engine was at work in the county in August 1716 which is close enough to Kalmeter's 1725 date of "about seven years ago" to at least allow one to entertain the suggestion that the "*London Gazette* engine" of August 1716 and Kalmeter's "Wheal an Vor" engine are one and the same. This would seem to rule out Carne's date of 1710 – 1714 (although see below). Significantly, both Carne and Kalmeter agree that the Wheal Vor engine was at work for around four years suggesting that they were probably talking about the same engine, albeit with Carne's dates becoming confused by the passage of time. Interestingly Thomas Newcomen spent the whole of 1718 in Cornwall, and part of this time may have been connected with the engine at Wheal Vor.[11]

In addition to the 1716 *London Gazette* reference there is at least one other piece of evidence dating from the 1710s: On 14th November 1717 George Liddell wrote to John Meres requesting that the Proprietors grant a licence to erect an engine at Farnacres Colliery in the North East of England.[12] On 30th November 1717 Meres replied agreeing to the terms suggested by Liddell. This letter is of fundamental importance as it contains the following statement:

"We hope we shall have a barrell and boiler for you in a short time, having some coming round from Cornwall which will fit your purpose........."[13]

If we accept that Kalmeter's "Wheal an Vor engine" was working between 1716 – 1720 / 21 this means that "November 1717 barrell (*i.e.* cylinder) and boiler" represent a second Cornish engine. This opens up the possibility that Carne's dates of 1710 – 1714 are in fact correct and the 1717 "barrell and boiler" originated from this engine. If this is the case Kalmeter's engine may have been its replacement. If this is not the case then one needs to consider that the 1717 "barrell and boiler" came from another engine; possibly the near mythical Balcoath turf burner.

If one is to draw any conclusions from the foregoing, one would have to concur with Allen and Rolt that it would be reasonable to assume that Newcomen had built a full size engine prior to the Dudley Castle engine. Similarly it is also reasonable to assume that, if such an engine existed, it might have been erected in Cornwall. There is indeed a "vague tradition", that there was a very early, turf burning engine erected at Balcoath Mine. One could speculate, and it is purely speculation, that if this engine did exist it was Newcomen's first full scale engine, possibly dating from 1710 – 1711 as suggested by Allen & Rolt and, as such, it would be a transitional design somewhere between the experimental models which he used to develop his ideas and the fully formed Dudley Castle engine of 1712. However given the lack of direct contemporary evidence the case, in the current author's opinion, remains at best "not proven". There is the possibility that a "pre Dudley Castle" Newcomen engine was erected at Wheal Vor in 1710. When considering this engine Carne's claim should be treated with some caution, although not necessarily dismissed. On the basis of the *London Gazette* and Kalmeter's notes one can argue with a reasonable degree of confidence that a Newcomen engine had been erected on Wheal Vor by August 1716 and worked for around four years until 1720 or 1721. At the current time this is the earliest Cornish engine for which we have anything approaching robust evidence. Nevertheless it would be a mistake totally to discount the possibility of earlier, possibly experimental, engines at Balcoath or Wheal Vor (or indeed elsewhere!) particularly in the light of the 1717 cylinder and boiler.

Myth, and legend, legend and myth.

For an interesting and in-depth examination of this subject, which draws significantly different conclusions from those drawn by the current author, the following article is highly recommended:

Greener J.,(2015), Thomas Newcomen and his Great Work, *Journal of the Trevithick Society*, No. 42, 2015, pp. 63 – 126.

Chapter 4 References

1. Cited in Allen J. S. & Rolt L. T. C., 1997
2. *Ibid*

3. *Ibid*
4. Allen J. S. & Rolt L. T. C., 1997
5. *Ibid*
6. Brooke J., 1993
7. *Ibid*
8. Carne J., 1828
9. Brooke J., 2001
10. *London Gazette* 11th August, 1716
11. Allen and Rolt, 1997
12. Cotesworth Papers, Gateshead Library, W/4/2 via James Greener
13. *Ibid*

Chapter 5
The 1720s to 1741

According to tradition we have to wait until 1720 before a Newcomen engine was erected on a copper mine in Cornwall. This engine is reputed to have been built by Thomas Newcomen and was said to have been erected on Wheal Fortune at Ludgvan. The earliest documentary source for this engine appears to be a paper read before the Statistical Society of London on 19th March 1838 by Sir Charles Lemon. He wrote:

> "The second steam – engine was erected at Wheal Fortune, in Ludgvan, in the year 1720....."[1]

Samuel Smiles, writing in the 1860s, offers the following details:

> "The first of Newcomen's construction in Cornwall was erected in 1720, at the Wheal Fortune tin mine, in the parish of Ludgvan, a few mile north – east of Penzance. The mine was conducted by Mr William Lemon, the founder of the fortunes of the well – known Cornish family..... The Wheal Fortune engine was on a larger scale than any that had yet been erected, the cylinder being 47 inches in diameter, making about fifteen strokes a minute. It drew about a hogshead of water at each stroke, from a pump 30 fathoms deep, through pit – barrels 15 inches in diameter, and its performances were on the whole regarded as very extraordinary."[2]

The 1720 Wheal Fortune engine seems to have become an accepted fact in spite of the fact that no contemporary sources support this view. A typical nineteenth century account may be found in the August 1868 issue of *One and All* which recorded that:

> "In 1720 Mr Thomas Newcomen came into Cornwall, on the invitation of Mr Lemon, and in the same year erected for him, at Wheal Fortune, a steam – pumping engine. This was the first engine seen on this principle in Cornwall. Newcomen's engine was so successful in freeing the mine of water, that the adventurers were able to work at a great depth and to cut into lodes which previously it had been impossible to explore".[3]

Such is the body of evidence regarding the 1720 Wheal Fortune engine, and it is interesting to note that Allen & Rolt accepted its existence. In a 1996 article in the *Journal of the Trevithick Society* Justin Brooke critically examined the evidence regarding this engine. Brooke was particularly sceptical of the nineteenth century accounts dismissing them as "folklore". In the article Brooke develops a strong case against the 1720 Wheal Fortune engine, part of which is that Kalmeter makes no mention of either the engine or the mine when he was in the vicinity in 1724, a strange omission if the mine was as successful as later commentators have suggested.[4] He offers a very plausible alternative hypothesis in regard to the Wheal Fortune engine. Brooke references William Penaluna's *The circle or historical survey of sixty parishes & towns in Cornwall* of 1819. Penaluna observes that the second engine was erected by Mr Lemon on Wheal Fortune in Ludgvan "about 70 years since".[5] If Penaluna is accurate this puts the date of the Wheal Fortune engine at around 1749 not 1720. Interestingly Thomas Goldney, acting as agent to the Coalbrookdale company, supplied engine parts to Joseph Percival and Co. at Ludgvan Lease in 1746. The parts supplied included a 47-inch diameter cylinder and bottom and a sinking pipe at a cost of £134.[6] The 1746 "Ludgvan Lez" engine is described in some detail in William Borlase's *Natural history* of 1758 and is discussed in the following chapter. Brooke notes that Smiles' description of the 1720 Wheal Fortune engine is lifted verbatim from Borlase's description of the 1746 Ludgvan Lease engine. Arguably the tradition

Ruins of Wheal Fortune. [By R. P. Leitch.]

A romantic nineteenth century view of the ruins of 1720 Wheal Fortune engine house by R. P. Leitch. This engine was wholly the creation of nineteenth century writers. It never existed in reality!

of the 1720 Wheal Fortune Newcomen engine derives from a misunderstanding on the part of Sir Charles Lemon regarding the 1746 Ludgvan Lease engine. This misunderstanding was perpetuated by nineteenth century writers and was accepted as fact by twentieth century writers with the notable and commendable exception of Justin Brooke.

The next Newcomen engines reputed to have been erected in Cornwall were a group of three engines dating from 1725 – 1727 said to have been the work of Joseph Hornblower (1696 – 1761), founder of a dynasty of steam engineers whom we will meet in later chapters. Like Thomas Newcomen, Joseph Hornblower was a devout Baptist which would have been a strong link between the two men. The surviving records of the Netherton Cinder Bank Baptist Chapel show that Joseph Hornblower was baptised there at the age of sixteen on the 24th August 1712. Cinder Bank is only around two miles as the crow flies from the location of the Dudley Castle engine and it is highly probable that Joseph Hornblower would have been fully aware of the new engine. He would certainly have known fellow members of the Cinder Bank congregation, Elias Newcomen and John Dunford who were in all probability engaged in erecting the engine. From here it is but a short step to establishing contact with Thomas Newcomen. Hornblower family tradition holds that the young Joseph Hornblower had worked in Cornwall with Thomas Newcomen.[7] As has been noted Newcomen spent the whole of 1718 in Cornwall and this might have been when Joseph worked with him, possibly in connection with the Wheal Vor engine.

Evidence for Joseph's activities in Cornwall in the 1720s comes from a book, published in 1863, written by his great grandson Cyrus Redding and entitled *Yesterday and Today*. This appears to be the key source for most later writers from Smiles onwards, the current author included. Given Redding's distance from the events he was describing a degree of caution is probably advisable as in cases of verifiable fact he is certainly not infallible. Referring to Joseph, Redding writes:

> "He had visited Cornwall for the purpose of erecting some of Mr Newcomen's engines at the mines about 1725......... It could not be many years after 1720, that the first engine was erected in Cornwall, near the North Downs at Huel Rose, seven or eight miles from Truro, and Mr. Joseph Hornblower, thus mentioned was the engineer, who had been sent for into Cornwall on purpose."[8]

Regarding the second Joseph Hornblower engine Redding cites the views of his cousin:

> "Mr Moyle, of Helston, an eminent medical practitioner there, and my first cousin, said in reply to a letter of mine on the above subject, in 1833:– "I think it is probable that the above engine was erected above a hundred years ago (referring to that at North Downs, Huel Rose, I believe). My uncle Mathew says he has often heard that the *second* engine was erected at Huel

Busy, or Chacewater mine, and that our great – grandfather Joseph was the engineer......"[9]

He concludes:

"After erecting these two engines it appears that my great - grandfather erected a third at Polgooth Mine. He then left the County entirely........"[10]

Thus if we accept Redding's account, and certainly more eminent historians than the current author do, three Newcomen engines were erected in Cornwall by Joseph Hornblower after 1725. Rolt & Allen offer the following dates for these engines: Wheal Rose 1725, Chacewater 1725 – 7 and Polgooth 1725 – 1727.[11] It is highly significant that the Costers / Bristol Company had interests in all three of these mines. This again demonstrates both the financial resources available to the Bristol Company and also John Coster (III)'s readiness to embrace new technology.

With regard to Wheal Rose, Kalmeter, who visited the mine on the 4th December 1724, notes that the mine was initially a rich tin mine, later becoming a "distinguished copper mine". In spite of its wealth the mine had not been worked for a number of years due to problems with water. Kalmeter records that there was a water wheel on the mine, however this obviously was not sufficient for the task in hand. In an attempt to dewater the mine the adventurers decided to drive an adit, work having commenced in around 1710. It was estimated that it would take a further three years *i.e.* 1727 to bring the adit home. If Redding's date of 1725 is correct the erection of an engine on Wheal Rose may well have been an attempt to bring the mine back into production before the adit reached the flooded workings; certainly it would be much safer to drain the flooded working using an engine than holing into "the house of water". That said the real strength of having a Newcomen engine is that it would allow working below adit and that the engine would really be of value when the adit had reached and drained the old workings.[12]

Both Chacewater and Polgooth have already been discussed at some length in connection with the Coster's water engines. The erection of a Newcomen engine at Chacewater marked the beginning of a history of engines on the mine which would, in microcosm, encompass the history of steam pumping engines in Cornwall. The introduction of Newcomen engines at Chacewater and Polgooth no doubt demonstrates that the Costers' patent water engine had reached its limit by the mid 1720s.

By 1733, when the extended Savery patent expired, the Newcomen engine had established itself in most British mining fields and indeed abroad; Allen and Rolt suggest that somewhere in the region of one hundred Newcomen engines had been erected.[13] However in Cornwall by 1733 the figure was possibly as few as four or five. This seems somewhat anomalous given the increasing depth that many Cornish mines were achieving, however there was one reason why adoption was not more widespread: Coal. Whilst blessed with seemingly endless mineral wealth, Devon and Cornwall are bereft of coal which had to be imported from the South Wales, Forest of Dean or Bristol coalfields; coal was an expensive commodity. Unfortunately for the mine adventurers in Devon and Cornwall, Newcomen engines had a voracious appetite for the stuff. This was a direct consequence of the thermally inefficient operating cycle of the engine which required the alternate heating and cooling of the cylinder, a problem not satisfactorily resolved until Watt's introduction of the separate condenser. Massive coal consumption was not a huge issue for engines erected on coal mines which had an almost inexhaustible supply of cheap or even free coal; indeed on some coal mines Newcomen engines continued in use into the late 19th or even early 20th century. However for the West Country mining man the cost of coal was of paramount importance. Writing in the 1770s William Pryce observed;

> "The vast consumption of fuel in those engines, is an immense drawback upon the profits of our Mines. It is a known fact, that every fire engine of magnitude consumes to the amount of three thousand pounds worth of coal in every year. This heavy tax upon Mining, in some respects, amounts to a prohibition*."[14]

The high cost of coal was compounded by an import tax on seaborne coals, apparently introduced in 1698 during the reign of William III to help meet the costs of the war of the English Succession (1688 – 97).[15] Pole in his *Treatise on the Cornish Pumping Engine* of 1844 notes that "the duty on coal alone required for one engine only amounted to nearly £350 per annum".[16] In response to the tax on sea borne coal a group of mine adventurers, foremost amongst whom was William Lemon prepared a "memorial" on the subject which was submitted to Parliament. Unfortunately the memorial is not dated but in all probability dates from the late 1730s:

> "The Case of the Gentlemen Adventurers in Tin and Copper Mines in the County of Cornwall, and the Inhabitants of the said County.
>
> For these many years past, there have been no new lodes or veins either of tin or copper discovered; and by the extraordinary and indefatigable labour of the miners, the county has been so entirely tried, that there is not the least reason to expect there will be any, so that unless some means can be found out to work the old mines at advantage, there is the greatest probability that the commodities of tin and copper will in a few years greatly decrease.
>
> This has obliged us humbly to desire Parliament to encourage fire engines by granting a draw back of the duties upon all coals consumed in working

***Pryce's estimate that a Newcomen engine consumed £3000 worth of coal per annum is a massive overestimate. In the mid 1750s engines at North Downs and Chacewater were consuming in the region of £300 worth of coal per annum.**

them, the only method we know of by which the old mines can be worked, the greatest and most considerable part of them being so deep that all other means for draining water out of them have already proved ineffectual.

The advantage is so apparent that we humbly hope this will meet with no opposition. It is evident the revenue cannot be lessened by it, because there are at present no coals at all consumed in that way; but on the other hand, if the mines remain unworked, the revenue must greatly suffer, since the duties on coals consumed in making tools and engine materials for working the mines, on iron, candles, gunpowder, foreign timber, and deals, great quantities of which are consumed in the mines, will be considerably diminished, as well as his Majesty's duties arising from tin exported.

It is therefore humbly desired, in consideration of the great expense the adventurers must necessarily be at, that this encouragement may be given, (as

A nineteenth century rendering by R. P. Leitch of a Newcomen engine house at Polgooth Mine.

well as a further encouragement by a draw – back of the duties on all coals used in the calcining of tin ore,) to preserve the said valuable commodities so very advantageous to the public, and without which thousands of families now employed in the several branches of these manufactures will be thrown out of business, and the greatest part of the county entirely ruined; or that such other methods be taken as Parliament in their wisdom shall think most meet."[17]

Of particular interest is the assertion that no engines were at work. Pryce, in considering the duty on seaborne coal, notes that "thirty – six years ago, this county had only one fire engine in it". Given that Pryce's book was largely completed by 1774/5 this places his statement at 1738 /39.[18]

Chapter 5 References

1. Cited in Burt R., 1969
2. Smiles S., 1865
3. Anon, 1868
4. Brooke J, 1996
5. Cited in Brooke J., 1996
6. Rogers K., 1976
7. www.penwood.famroots.org
8. Redding C., 1863
9. *Ibid*
10. *Ibid*
11. Allen J. S. & Rolt L. T. C., 1997
12. Brooke J., 2001
13. Allen J. S. & Rolt L. T. C., 1997
14. Pryce W., 1778
15. Howard B., 1999, Pole W., 1844
16. Pole W., 1844
17. *Ibid*
18. Pryce W, 1778

Chapter 6
The Newcomen engine in Cornwall after 1741

The memorial outlining "The Case of the Gentlemen Adventurers in Tin and Copper Mines in the County of Cornwall, and the Inhabitants of the said County" appears to have been successful for in 1741 Parliament passed the remarkably titled *"Act for granting to His Majesty the sum of One Million out of the Sinking Fund, and for applying other sums therein mentioned for the service of the year one thousand seven hundred and forty – one; and for allowing a Draw – back of the duties upon Coals used in Fire Engines for draining Tin and Copper Mines in the County of Cornwall; and for appropriating the supplies granted in this session of Parliament; and for making forth Duplicates of Exchequer Bills, Lottery tickets and Orders, lost, burnt otherwise destroyed; and for giving further time for the payment of duties omitted to be paid for the Indentures and Contracts of Clerks and Apprentices*".[1]

To reclaim the 1741 drawback representatives of the mines operating an engine were required to swear an oath to the Customs authority in the port where the duty had been paid as to how much coal had been consumed.

In considering the granting of the 1741 drawback Bridget Howard observes that it was not granted out of any concern for the Cornish mining industry but, rather, was a consequence of cynical political expediency on the part of the Whig prime minister Robert Walpole (1676 – 1745) who was fighting for political survival. During the 1730s Walpole's star appears to have been on the wane; whilst maintaining his hold on power at the General election of 1734 Walpole had lost 85 seats.[2] At the 1741 election one of Walpole's key opponents' was the Prince of Wales who was also the Duke of Cornwall and, consequently, had significant influence within the county. In an attempt to sway the 44 MPs who sat in Cornish constituencies Walpole offered them what was in effect a bribe in the form of the drawback. Whilst Walpole did win the 1741 election he did so with a very small majority and his premiership lasted only until February 1742 when he was forced to resign after being defeated in a motion of no confidence.[3]

If the abolition of the duty on seaborne coal failed to bolster Walpole's failing political career contemporary commentators believed that it had a significant effect on the adoption of Newcomen engines in Cornwall; Pryce comments:

"The drawback upon coal used in our smelting houses and fire engines, has been attended with such happy consequences for the publick, that we may venture to affirm, not one – fifth of the fire steam engines now working, would ever have been erected without such encouragement. Thirty – six years ago, this county only had one fire engine in it: since which time above three score have been erected, and more than half of them have been rebuilt, or enlarged in the diameter of their cylindrical dimensions."[4]

William Borlase, in his *Natural History of Cornwall* of 1758, similarly notes that: "..... without this bounty, fire engines would not have been erected, nor could these mines ..... ever have been worked....."[5] Given the perceived significance of the drawback one would expect to find a flurry of engine building in immediate aftermath of the 1741 Act. This certainly seems to be Barton's thinking on the subject as he writes in his *Cornish Beam Engine* of 1966:

"In 1742 alone the Coalbrookdale Company supplied Hornblower senior no less than five cylinders which were quite probably intended for engines on William Lemon's copper mines in Gwennap and elsewhere."[6]

Unfortunately, although not atypically, Barton does not cite his source for this statement and it has not proved possible to verify his assertions; certainly no reliable historian makes reference to these cylinders. Verifiable documentary evidence of cylinders supplied by Coalbrookdale may be found in the surviving records of the company. Mott in his 1963 paper for the Newcomen Society examined these records which do not include Barton's five 1742 cylinders, nor indeed anything which might be construed as such.[7] Details of engine parts supplied to Cornish mines by the Coalbrookdale Company in the post drawback period may be found in the account books of Thomas Goldney (1694 – 1768). Goldney was a Bristol merchant and banker, who held three sixteenths of the shares in the Coalbrookdale Company, and who acted as agent for the company in the South West. During this period the Coalbrookdale company dominated the manufacture of cast iron cylinders for Newcomen engines. The account books cover the period 1741/ 2 to 1769 have been examined by K. H. Rogers. Like Mott, Rogers found no evidence of the five 1742 cylinders.[8]

The earliest record of a "Cornish cylinder" found by Rogers in Goldney's account books is for the delivery, on 31st December 1744, of a bored cylinder weighing 43 hundredweight, a bottom and five barrels for Abel Angove of "Trevenson in the Parish of Logan" (Illogan). The cylinder was found to be faulty and a replacement was despatched from Coalbrookdale in October 1745. By the end of 1746 Goldney had supplied a further four Coalbrookdale cylinders to Cornish mines viz: Ludgvan Lease (47″ cylinder), Roskear (47″ cylinder), Dolcoath (40″ cylinder), Lemon & Co. (mine unspecified; 40″ cylinder, estimated).[9]

One needs to ask the question as to why there was an apparent lull in engine building between 1741 and 1745 / 1746? John Rowe, in his *Cornwall in the age of the industrial revolution,* makes an interesting suggestion. He considers that the duty on sea borne coal was only one constraint to the widespread adoption of the Newcomen engine in Cornwall. He suggests that the lack of skilled engineers in Cornwall was also a significant handicap to the post 1741 adoption of the Newcomen engine.[10] This is debatable; adventurers in Cornish mines were quite capable of importing skills from beyond the boundaries of the County as is attested by the employment of Joseph Hornblower by the Costers in the 1720s. Arguably if there was a demand for Newcomen engines during 1741 – 1745 it would not have been beyond the wit of Cornish mine adventurers to find the necessary skills which certainly did exist within Britain. It is significant that the 1746 upsurge in engine building coincided with the arrival of Jonathan Hornblower (I) (1717 – 1780) in Cornwall. Jonathan was the son of Joseph Hornblower and was a skilled erector of Newcomen engines in his own right, having gained experience erecting engines with his father in Derbyshire, Shropshire and Wales. Jonathan Hornblower and his wife Ann left Broseley in the Midlands on 22nd August 1745 arriving at Truro on 6th October 1745.[11] It would be surprising if there was not a significant correlation between the arrival of Jonathan Hornblower and the expansion of engine building in Cornwall. It is probable that Hornblower's arrival in Cornwall was a response to an emergent demand for engines and not the other way around, as implied by Rowe.

To get to the heart of the matter it is worth examining the growth of copper exports from Cornwall during the 1740s.

Table 1: Cornish copper ore exports 1740 – 1749, after Morton J., 1985[12]

Date	1740	1741	1742	1743	1744	1745	1746	1747	1748	1749
Tons	4700	5968	5751	6826	7393	9061	7900*	10000*	8787	6278

*Figures for 1746 & 1747 from Tann J., 1996.[13]

What springs out is the spike in Cornish copper production in 1745 – 8. This "boom" coincides very nicely with the arrival of Jonathan Hornblower in the county and the upsurge of engine building. A much more satisfactory explanation than Rowe's is that an increasing demand for Cornish copper ore, possibly a partial consequence of the War of the Austrian succession (1740 – 1748), led to increased output from Cornish mines. Increased output meant that mines had to go deeper, exacerbating the problems of pumping, one solution to this being the Newcomen engine. The 1745 boom represents the point where the application of a Newcomen engine becomes economically viable and thus attractive to mine adventurers. A demand for such engines having arisen it is not surprising that an engine erector of Jonathan Hornblower (I)'s ability was attracted to the county, especially when one considers his father's earlier association with Cornwall. If one accepts this scenario it is but a short step to accepting that it was the 1745 copper boom which kick started the widespread adoption of the Newcomen engine in Cornwall and not the 1741 draw back as has traditionally been thought. That said, one should not ignore the importance of the 1741 draw back which lowered the threshold where it became economically viable to employ a Newcomen engine.

Reinhold Rucker Angerstein, yet another Swede, visited Cornwall in the early summer of 1754 and left a record of his observations. Angerstein recorded details of eight engines, seven of which were at work. Apart from their intrinsic interest Angerstein's observations are of value as it is possible to correlate them with details of Coalbrookdale cylinders supplied via Goldney.

On the 15th May Angerstein visited Polgooth where he noted that Mr Lemon had installed two steam engines.[14] The Goldney papers record that a Coalbrookdale cast cylinder of around 40″ was supplied to Polgooth in 1747. The second engine may also have been around 40″ as in 1746 William Lemon & Co. purchased a cylinder of this size, again via Goldney.[15] Angerstein records that these engines were not adequate for the job in hand:

> "they do not have sufficient capacity to take all the water away, and the work in the mines has to stop for a couple of months at the wettest period."[16]

On the 20th May 1754 Angerstein travelled to Redruth from where he visited North Downs, commenting:

> "Most of it was now flooded, which generally happens in wintertime. They were, nonetheless, busy with the installation of a new fire engine in addition to the old one and the water engines erected underground in the adits."[17]

Angerstein reports that the new engine at North Downs had a cylinder 69″ in diameter. In all probability 69″ inches is an error; Goldney supplied a 60″ Coalbrookdale cylinder to Wheal Rose in 1753 and this, no doubt, is what Angerstein saw.[18] The boiler was fifteen feet in diameter, thirty feet high; it was constructed from iron plates and weighed six tons. It seems to have been the practice during this period for engines (both water and fire) to be supplied by an external contractor. This was certainly the case with the new engine at North Downs where "the partners pay the contractors for the fire engine one-fifth of all ore raised during the first seven years. Thereafter the engine will belong to the partnership without further payments".[19] Staying in the North Downs area Angerstein noted that a fire engine at Truan Tin Mine was raising eight to ten hogsheads of water per minute. From Redruth Angerstein moved on to Pool where shafts had reached a depth of two hundred and seventy feet. The mine, worked by Squire Praed and Abel Angove, was drained by a fire engine and an adit at the "150 – feet level". The Pool engine had a 60″ Coalbrookdale cylinder supplied in 1748.[20] Dolcoath was drained by a fire engine and three water wheels. By the time of Angerstein's visit Dolcoath had received two Coalbrookdale cylinders, a 40″ supplied in 1746 and a 54″ supplied in 1753. Given that Angerstein reports only one engine on the mine one could speculate that the 54″ replaced the 40″.[21] In common with other commentators Angerstein notes that Chacewater was a wet mine. At the time of his visit Chacewater was worked by a fire engine and water engines, the latter possibly built by the Costers:

> "There are two water – engines in one shaft here, that are barely sufficient to keep the mine free from water, especially during the winter, because it is low - lying and attracts water from other smaller mines between Chacewater and North Downs. These are worked by other owners who also enjoy the benefits of the fire engine without contributing anything to it."[22]

The engine at Chacewater would have had a 54″ Coalbrookdale cylinder supplied in 1750.[23]

Borlase's *Natural History* of 1758 is probably the best known and most frequently quoted (and indeed misquoted *vide* Barton[24]) source for Cornish Newcomen engines of this period. Borlase notes Newcomen engines at the following mines: Bosprousal, Bullen Garden, Dolcoath, Herland (Drennack) Mine at Gwinnear (70-inch cylinder), Ludgvan Lease work in Ludgvan (47-inch cylinder), North Downs (two engines), Pittslooarn and Metal Works (two engines) (*Chacewater*), Polgooth, Pool, Wheal Reeth in Godolphin Bal and Wheal Rose. The list is not exhaustive, Borlase observes that, in addition to the mines he mentions, "some others" were also worked by fire engines.[25]

Borlase's informants regarding Newcomen engines included both William Lemon and John Nancarrow. Quite remarkably some of Lemon's and Nancarrow's correspondence with Borlase survives. Borlase's letter books are currently in the care of the Morrab Library in Penzance.

In a letter to Borlase dated 8th May 1756 Lemon provides some details of Newcomen engines on his mines. Two engines were in operation at North Downs and, on a monthly basis, were consuming 199 "Chaldns", 4 weys of coal at five shillings per "chalder" or £49 15s per month. Extrapolating this, both North Downs engines were consuming £597 worth of coal per year, or £298 10s each. This figure is somewhat at odds with Pryce's assertion "that every fire engine of magnitude consumes to the amount of three thousand pounds worth of coal in every year".[26] Lemon also makes reference to "Pittslewerne" and Metal Work where two engines were in operation consuming 248 "Chaldrns", 32 weys of coal per month. The coal consumption of the latter pair of engines at Chacewater is somewhat heavier than the pair of engines at North Downs; this is understandable given that "Pittslewerne" and Metal Work were considered to be amongst the wettest mines in the county. Lemon also mentions an engine at Polgooth consuming 111 chaldrons, 24 weys per month.[27]

William Borlase's letter books contain a very interesting account, dating from 1757, written by John Nancarrow in the form of a series of questions and answers. Given the historical interest of the document Nancarrow's letter is reproduced as transcribed:

"Query 1 How many inches diam: the cylinder of the Great Work?
Answer 47 inches
Qu: 2 How many inch diam is the cylinder at Herland?
Answer 70 inches
Qu: 3 How many tons of Water does the G Work Engine draw p Minute?
This depends on the depths and the Diam of the Working Barrells (or Cylinders) in the Shaft. The House Cylinder was brot from Ludgvan to G. Work; there we drew abt 30 fathoms under the addit; The pit Barrells were 15 inches diam & the water delivered to the addit was abt an Hogshead an Stroke, and generally went about 14 or 15 Strokes p minute – At Great Work our Pit Barrells are but 10 Inches Diam: and is calculated for drawing between fifty and sixty fathoms under the addit, which addit is about 30 fathoms deep. As the quantity of water raised in a given time is the squares of the Diameter of the pit Barrell the same cylinder does not lift half the quantity here as at Ludgvan Lease. We also raise the water from the addit with a six inch box to feed the boilers & c.
Query 4th How many Tons p minute doth the Mannour (*Author's note: mine not known)* engine draw?
Answer Cannot say, as I don't know the Diam of the pit Barrell nor the depth drawn.
Query 5th When the Engine begins to beat, or strike on the springs, is it not occasioned by the Violence of the Steam & from other concerning circumstances. And whether when the motion of the Engine is become so violent to beat, it does not let fall the Damper and by that one stop itself.
Answer, The Steam when strong hath some small effect by Ent'ring (?) the throat pipes in filling the Cylinder in the short (?) space but as the Engine is fixed to a certain stroke from spring to Spring, we have several methods for lengthening and shortening the stroke; and a sort of Trigger that falls into a Notch made in the F (Just as the Engine begins to strike the springs) which prevents it from making another stroke by stopping the Injection Cock; for of course if the injection is stopped there can be no vacuum and the Machine must stand until somebody sets it on again.*
The Damper is of great use in regulating the fire, but never used as the immediate cause of stopping the engine for that only causes the fire to slacken slowly, whereas the engine when it begins to beat must be stopped in an instant or great damage must ensue. If the steam becomes too strong we presently

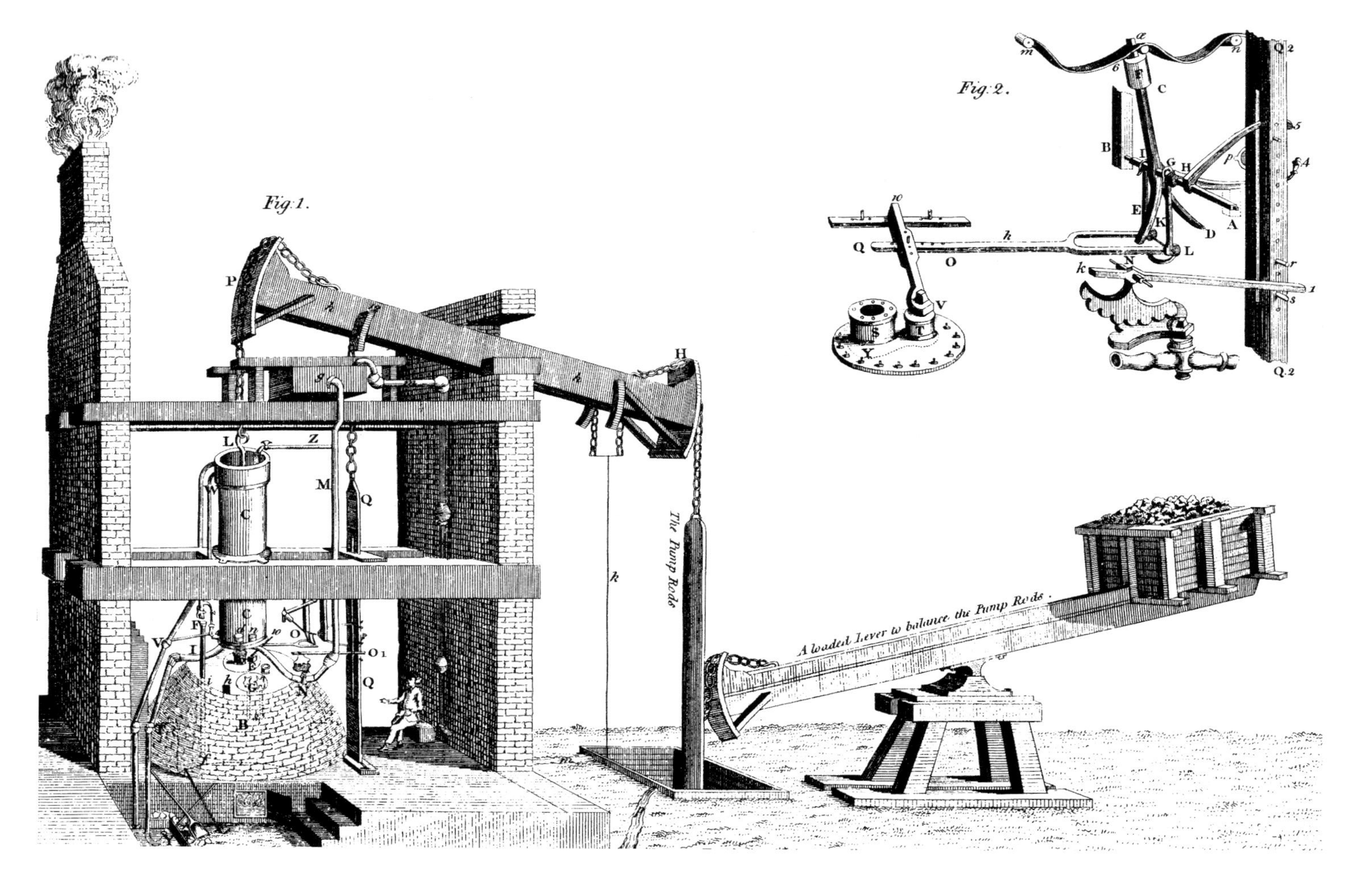

Engraving of an atmospheric engine from Pryce's *Mineralogia Cornubiensis* of 1778. Francis Trevithick writing in his "Life of Richard Trevithick" of 1872 "suggests" that this engine is in fact a 45″ engine erected by Richard Trevithick Senior. *Image courtesy of Tony Clarke.*

discharge it through the steam pipe contrived for that purpose and is large enough to carry off all the steam, even when the Engine stands.

Query 6 For whom was the Old Mr Costar agent for buying Copper Ore?

Answer I Do not know.

*The greatest and most frequent cause of the Engines beating (or as the North countrymen call banging) is, when the water is in fork and the air gets in with the water on to the pumps, The Engine then wants her due load, raises lighter & quicker, also returns quicker; the zed or sweeper acting with greater force on the outer end of the Beam, the resistance of the Rod, met with in descending through the pumps being lessened in proportion as there happens to be more or less air mixed with the water.

I sincerely wish thee a good journey & safe return I am with great respect thy obliged Fr
John Nancarrow."

As the reader will appreciate, sources for Newcomen engines in Cornwall are somewhat fragmentary. One source that has so far been neglected by students of this subject are the Customs records. For a mine to claim the drawback it was necessary to swear an oath to the customs authority who duly recorded the details. Thomas Wilson, Boulton and Watt's Cornish agent, cites examples from the Custom's records in a pamphlet entitled *A comparative statement of the effects of Messrs Boulton and Watt's steam engines with Newcomen's and Mr. Hornblower's* of 1792:

"I may now state from authentic Documents to which reference may be had: that upon examining the Custom – house books at Truro, a Debenture or Drawback on Coals, was paid to the Adventurers of Tresavean Mine for 1917 Chaldrons of Coals of 36 Bushels each, from 31st July 1768 to August 1st 1770 being two Years, which being reduced into Weys of 64 B (ushels) is 1078 W. 12. 4B or 539 W. 02. 10B. in one Year, consumed on that Mine by one Engine of the Old Construction.

By the Custom – house Books at Truro and Penryn it appears that a Drawback was paid to the Mines of Wheal Virgin, Wheal Maid and West Wheal Virgin, from June 30th 1778 to July 1st 1779, for 6362 W. 32. 12B of 64 B. consumed by seven engines of the old sort then working on these Mines.....

I procured from St. Ives the consumption of two old engines working upon the Wheal Chance Mine, in Camborn; these were esteemed very good ones, ... from 3rd Octob. 1777 to October the 3rd 1778, their consumption was 1093w 0q 10b of 72b each."[28]

Although beyond the resources of the current work, a thorough study of surviving Customs records would undoubtedly provide extremely valuable insights into the subject.

One of the last, if not the last, Newcomen engines to be erected in Cornwall was the "Dolcoath New Engine" of 1775. The term "new" was something of a misnomer as the engine had been acquired from Carloose Mine. The engine had a 45″ cylinder and was erected by John Budge.[29] Fortunately details of the costs associated with the engine have survived:[30]

Dolcoath New Engine Cost Account.

Carloose adventurers for materials	£414	12s	3d
John Commins, for boiler top, &c	£93	8s	9d
John Jones & Co. (Bristol), for iron pumps	£118	6s	10d
Dale Company, for iron pumps	£131	9s	4d
Mr Budge, for erecting the engine	£63	0s	0d
Carriage of the boiler, cylinder, &c., from Carloose including attendance, &c., &c.	£50	0s	0d
Arthur Woolf, per month	£1	14s	4d
John Harvey and partners, for putting in the boiler and building shed - walls, &c.	£33	1s	9d
To new ironwork, as per account	£187	10s	4d
To timber boards, &c., as per account	£255	10s	10d

Chapter 6 References

1. Cited in Pole W., 1844
2. Wikipedia, British general election, 1734
3. Wikipedia, British General election 1742; Howard B., 1999
4. Pryce W., 1778
5. Borlase W., 1758
6. Barton D. B., 1966
7. Mott R. A, 1964
8. Rogers K. H., 1976
9. *Ibid*
10. Rowe J., 1993
11. Harris T. R., 1976
12. Morton J., 1985
13. Tann J., 1996
14. Berg T. & P., 2001
15. Rogers K. H., 1976
16. Berg T. & P., 2001
17. *Ibid*
18. Rogers K. H., 1976

19. Berg T. & P., 2001
20. Rogers K. H., 1976
21. Berg T. & P., 2001; Rogers K. H., 1976
22. Berg T. & P., 2001
23. Rogers K. H., 1976
24. Barton D. B., 1966
25. Borlase W., 1758
26. Pryce W., 1778
27. Borlase Letter Books
28. Wilson T., 1792
29. Trevithick, F, 1872
30. Cited in Trevithick F., 1872

Chapter 7
Pushing the limits of the Newcomen engine

The post 1741 history of the Newcomen engine in Cornwall is dominated by two themes. First there was a desire to extract more and more power from Newcomen's original design reflecting the increasing depths of the County's mines and secondly, there was the ongoing concern regarding the fuel consumption of such engines.

The most obvious way to increase the power of the Newcomen engine was to increase cylinder size. Borlase writes:

> A cylinder of forty – seven inches at Ludgvan – lez work – in the parish of Ludgvan, making about fifteen strokes a minute, usually drew through pit – barrels of fifteen inches diameter, from a pump thirty fathoms deep, about and hogshead at each stroke, that is fifteen hogsheads of water in each minute......... But the cylinders can be made much larger, that at Herland (or Drennack) mine, in the parish of Gwinear, is seventy inches in diameter, and will draw a greater stream of water at any equal depth....... The only objections to this engine are the great expenses in erecting, and vast consumption of coals in working it."[1]

The list of Coalbrookdale cylinders supplied by Thomas Goldney to Cornish mines between 1744 and 1768 demonstrates a rise of cylinder size throughout this period:[2]

1744:	Trevenson (40″ estimated)
1745:	Trevenson (40″ estimated), (replacement for the 1744 cylinder)
1746:	Ludgvan Lease (47″), Roskear (47″), Dolcoath (40″), Lemon & Co. (mine unspecified) (40″ estimated)
1747:	Polgooth (40″ estimated)
1748:	Pool Adit (60″), Drannack (55″)
1749:	Lemon & Co. (Mine unspecified) (52″)
1750:	Chacewater (54″)
1753:	Dolcoath (54″), Herland (70″), Wheal Rose (60″)
1756:	North Downs (60″)
1758:	Wheal Virgin (60″)
1763:	Wheal Oula (70″), Poldice (60″)

Engraving by Francois Vivares of a Newcomen engine cylinder on a wagon leaving Abraham Darby's Upper Works at Coalbrookdale, circa 1758. The image highlights the problems of moving large engine components at the dawn of the industrial revolution. A cylinder bound for Cornwall would start its journey by horse drawn wagon; it would then be loaded onto a trow (barge) for its journey down the River Severn, being transferred to a seagoing vessel for its journey to Cornwall before being transported to the mine by wagon and horses. ***Copyright of and courtesy of the Ironbridge Gorge Museum Trust – the Sir Arthur Elton Collection – AE185.769.***

1765: Wheal Virgin (60″)
1765: Tresavean (60″)
1766: Chacewater (66″)
1767: Wheal Virgin (60″)
1768: New Dolcoath (63″)

Increases in cylinder size were made possible by improved casting and machining techniques. In this context Allen and Rolt observe that:

> "The development of Newcomen's engine in the eighteenth century must be seen against this background of slow but certain improvement in workshop practice".[3]

An alternative to ever greater cylinder size was the use of using multiple engines pumping the same shaft as was the case at Chacewater:

> "There were formerly two atmospheric engines working on this mine; one a 64 inch cylinder, the other a 62 inch; both six feet stroke. These two engines were stated to consume 16½ bushels of coal and hour. The quantity of water they raised to keep the mines drained, was 80 cubic feet per minute, in summer and 100 cub. ft. per min. in winter.
>
> One of these engines worked the lower column of pumps, 18½ inc. diam., and 24 fathoms lift, and the other engine worked the upper column 17½ inc. diam. and 26 fathoms lift; so that one engine raised the water to the other, and that engine raised it up high enough to run away, by the subterranean level or adit, which was 24 fathoms below surface. In the whole the water was raised 50 fathoms.
>
> This was a common arrangement for draining the large mines in Cornwall, which required more power than one engine usually possessed; this plan was called shammeling."[4]

It is probable that the 60″ and one of the 66″ engines at Poldice were also shammeling, as were the Bullen Garden engines illustrated in Pryce's *Mineralogia Cornubiensis*.[5]

Whilst increasing cylinder size and shammeling were the obvious solutions to the challenge of increasing the depth which could effectively be pumped; there was the possibility of a third option, hinted at as early as 1758 by Borlase who refers to "increasing the elasticity of steam". In all probability the idea, as recounted by Borlase originated with John Nancarrow.[6] "Increasing the elasticity of steam" may be interpreted as the expansive use of steam with the possible implication of higher working pressures.

Section of Bullen Garden Mine at Camborne from Pryce's *Mineralogia Cornubiensis* of 1778 although the section is believed to date from the 1760s. The engine shaft is pumped by two Newcomen engines, a practice that became increasingly common as mines went increasingly deeper as the eighteenth century progressed. Also shown are a pair of "bob engines". The conical structures housed whims. *Image courtesy Tony Clarke.*

Fuel consumption was never far from the forefront of Cornish adventurers' minds and any innovations which might contribute to fuel economy were welcomed with open arms. One such was Sampson Swaine's proposal to employ waste heat from smelting furnaces to heat boiler water. Swaine's invention was covered by patent No. 774 granted on May 21st 1762. This patent actually covered two inventions: The use of waste heat from smelting to heat a boiler and, secondly, a device to derive rotary motion from a Newcomen engine:

> "New Method of constructing and adapting to each other a machine furnace and fire engine, so that the same fire should at the same time be capable of smelting and refining several sorts of metals and working a fire engine to raise water, to stamp ores, and serve many other useful purposes in a manner not hitherto put into practice."[7]

Soon after being granted the patent Swaine was able to put his ideas into practice. T. R. Harris notes that on December 14th 1762 Sampson Swaine and John Weston acquired a sett from Francis Basset on the course of the Wheal Weeth Main Lode. The sett grant permitted Swaine and Weston to "...... erect their new invented Engine with as many Furnaces as may be wanted to work the same...."[8] The grant required Swaine and Weston to erect a fire engine to dewater the mine within eighteen months. The engine, and presumably Swaine's patent boiler / furnace arrangement, was at work by September 1764.[9] Indeed it would appear that more than one of these devices was erected: The Frenchman Gabriel Jars in his *Voyages metallurgiques* records that in around 1768 examples had been erected at Wheal Kitty and Wheal Chance, adjacent to Wheal Weeth.[10] Regarding the Wheal Chance example Jars provides the following description of the furnaces:

> "In an emplacement excavated below the level of ground 22 to 24 feet long and 14 to 15 feet wide and underneath the boiler, which has the same dimensions, are constructed two reverberatory furnaces in one of which the metal is roasted and in the other is melted; above these is a third (furnace) of the whole length of the said boiler the arch (vault) of which is made of brick of a single thickness; but since the bottom of the boiler does not form the same arc as the said arch there is necessarily a void between them. In the angles of the furnaces there are small boilers which by their correspondence which they have with the large one make a single one only; independent of these three furnaces there is another smaller one placed at one of the angles but outside the boiler while three tubes of sheet iron about 1 foot in diameter traverse the whole length; each of these tubes receives the gasses and the flame which proceed from each furnace of cast iron and which transmit it to three chimneys placed at the other extremity. As to those which come from the roasting furnaces they are conducted the length of one side of the boiler which firing four different fires;

Moorstone boiler block on display outside the engine house at East Pool and Wheal Agar Mine (operated by the National Trust as 'East Pool Mine') at Pool. Behind this is a piece of cast-iron beam from an unknown engine. *Image courtesy Pete Joseph.*

a fifth which heats the other side is to be reckoned."[11]

Concentrating primarily on the arrangements of the furnaces Jars provides little detail of the boiler itself. Fortunately Jabez Carter Hornblower, writing in 1801 has left a description of one of Swaine's boilers:

> "It had long been a desideratum in the mining interest in that county, to reduce the consumption of fuel in draining the mines by means of steam engines, and every expedient that carried any tolerable face of probability was bought to a trial.
>
> Among the rest it was suggested, that as, in the several operations of smelting the produce of the mines, much heat must be carried off, from the intense fires of their furnaces, which might be employed to some purposes requiring but a subordinate; a resolution was formed by a company of gentlemen, with Mr John Weston at the head of it (to whom mining was an ungovernable hobby horse), to erect an engine on a copper mine in the parish of Camborne, and to put up a set of furnaces so attached to the engine that they might avail themselves of this superabundant heat to raise steam.
>
> To affect this they had their engine – boiler made of masonry, of what is called in Cornwall *moor – stone*, well wrought and put together with Aberthaw lime,

Sampson Swaine's moorstone boiler as re-erected at Dolcoath by Arthur Woolf senior in the early 1770s. ***Image courtesy Pete Joseph.***

which has the property of setting like the Dutch trass (*a volcanic tuff occurring in the Eifel, where it was worked for hydraulic mortar – Wikipedia*); and to convey the heat, two or three copper tubes were placed in it from end to end, and the furnaces connected to one end of the boiler, and the engine at the other [12] Farey adds further details of Swaine's boiler:

"By some notes sent to Mr Smeaton in 1773, it appears that this stone boiler was 20 feet long, by 9 feet wide within, and 8½ feet deep, the fire was applied in three copper tubes 22 inches diameter fixed within the water, and extending through all the length of the boiler. Two of these tubes were fixed within 7½ inches of the bottom of the boiler, and the other one was over the space between them, a 2 feet above the bottom. The flame was conducted from the furnace through one tube, then turned back into the other tube, and then returned through the third, as to pass three times through the whole length of the water."[13]

To modern eyes a moorstone or granite boiler appears bizarre in the extreme, however in the context of the times, with very low working pressures and the comparatively primitive state of metal working, a stone boiler was not an unreasonable proposition. Indeed after Wheal Weeth closed the remains of the moorstone boiler were sold in July 1771 to Dolcoath for £70 7s 3d where it was re erected, albeit without the smelting furnaces, in the vicinity of Old Sump Shaft probably by Arthur Woolf Senior.[14] Whilst the intentions which underlay Swaine's invention were admirable the execution may have been less good. Hornblower, for example, damns it with faint praise, commenting:

"..... it was a prodigal way of saving heat, yet it was competent enough to raise steam for the use of the engine in certain cases."[15]

A particular criticism, lev-

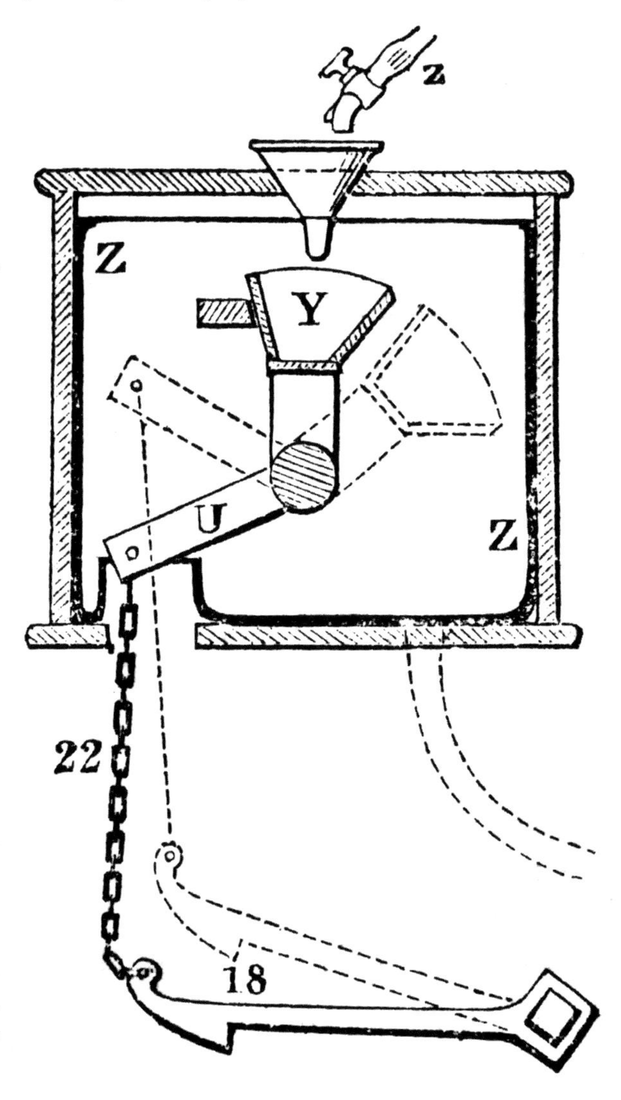

Illustration of cataract governor from John Farey's Treatise on the Steam engine of 1827. Water from pipe z (which is fed from the injector pipe) fills the container Y, the rate of filling being varied by the tap. When Y is full the "flop jack" tilts on its axle, the chain (22) actuating the lever (18). The lever (18) controls the injection valve which initiates the working stroke of the engine. Thus by controlling the flow of water into the cataract the number of strokes the engine makes can be controlled. The ability to control the strokes the engine made was very important in promoting efficient use of fuel, a priority in Cornwall. *Image courtesy Tony Clarke.*

elled by Jars, was that the furnaces required irregular firing whilst the boiler required a constant head of steam.[16] Farey notes that some contemporaries felt that too much heat was lost to "the great mass of stone work".[17] However in an environment where coal consumption was paramount it is not surprising that Swaine's ideas regarding the reuse of waste heat from furnaces would be revisited. John Smeaton thought the principle was worth trying with his 72″ Newcomen engine erected at Chacewater Mine in 1775.

The cataract was a significant development which probably originated in Cornwall; Farey noting that the devices were commonly in use in Cornwall by the 1770s. The purpose of a cataract was to:

> "regulate the motion of the engine to any given number of strokes per minute, so that the desired quantity of water could be drawn without wasting steam in working quicker than necessary."
>
> The cataract comprises of a "flop jack" mechanism. Water is fed from the injection pipe, via a tap, into the flop jack. When the flop jack is filled it tips over and empties. A chain is linked from the flop jack to the injection cock; thus when the flop jack tips it opens the cock allowing injection water to enter the cylinder, causing the engine to make a stroke. Having tipped and emptied the flop jack will return to the starting position and will start to refill. By altering the amount of water feeding into the flop jack you alter the speed it fills and tips and, hence the number of strokes then engine makes.[18]

Chapter 7 References

1. Borlase W., 1758
2. Rogers K., 1976
3. Allen J. S. & Rolt L. T. C., 1997
4. Farey J., 1827
5. *Ibid*; Pryce W., 1778
6. Borlase W., 1758
7. Patent Office 1867
8. Cited in Harris T. R., 1945
9. Harris T. R., 1945
10. *Ibid*; Jars G., 1823
11. Cited in Harris T. R., 1945
12. Hornblower J. C., 1801
13. Farey, J., 1827
14. Carter C., 1998
15. Hornblower J. C., 1801
16. Harris T. R., 1945; Jars G., 1823
17. Farey J., 1827
18. *Ibid*

Chapter 8
John Smeaton and the Chacewater 72″

It is not surprising that the polymath engineering genius John Smeaton (1724 – 92) should have turned his attention to the Newcomen engine. Unlike the majority of his contemporaries Smeaton brought scientific rigour to all his endeavours; his work on waterwheels[1] providing the template for his later work on Newcomen engines.

Smeaton commenced work on Newcomen engines in the mid 1760s, designing an engine for the New River Company in London in 1767. Smeaton used this engine as something of a test bed for his developing ideas. His intention was to run the engine slowly with large pumps (eighteen inch diameter with a thirty five foot lift) and to load the piston as much as was possible (10¼ lbs per inch). The beam had an off centre fulcrum meaning that a nine foot piston stroke delivered a six foot pumping stroke. The cylinder was only eighteen inches in diameter; Smeaton felt that the small diameter would be compensated for by the fact that most of the steam in the cylinder would be in contact with the cylinder walls causing rapid condensation. Smeaton commented that "I thought myself quite secure of the result". Whilst the engine performed its allotted task as intended Smeaton was far from satisfied with its performance; the engine requiring significantly more coal and injection water than Smeaton had anticipated. By altering the fulcrum position on the beam he managed to reduce the load on the piston from 10¼ lbs to 8¼ lbs per inch. Whilst this shortened the stroke, the engine ran faster and used less coal and injection water. Smeaton observed:

> "In short a single alteration seemed to have unfettered the engine; but the extent of this condensation under different circumstances of heat, and where to strike such a medium, as to obtain the best result, was still unknown to me. I resolved, if possible, to make myself master of the subject and immediately began to build a small fire – engine, which I could easily convert into different shapes for experiments."[2]

Smeaton duly built himself an experimental model at his home in Austhorpe in Yorkshire in the winter of 1769. This engine had a cylinder diameter of 10 inches and a working stroke of three feet, the engine made 17½ strokes per minute. Over a period of four years Smeaton carried out more than one hundred and thirty experiments. During each experiment Smeaton monitored and recorded both the mechanical

John Smeaton (1724 – 1792). Polymath engineering genius who designed some of the finest Newcomen engines ever built including the Chacewater 72″. On the left is Smeaton's Tower, the former Eddystone lighthouse. This was largely dismantled and rebuilt as a memorial at Plymouth Hoe.

power of the engine and also the power generated during the consumption of a bushel of coal. From this second measure Smeaton was able to develop the concept of duty which allowed the comparison of differing engines and would contribute significantly to the rapid competitive development of the Cornish engine in the first half of the nineteenth century. In carrying out his experiments it was Smeaton's practice to get the engine into the best working order, then alter one aspect of its operation and monitor the effect it had on overall performance. His experimental work allowed him to develop a set of principles relating to matters such as firing that he could apply to larger machines.[3]

Smeaton supplemented his experimental work by gathering empirical data regarding existing engines. In 1769, probably at Smeaton's instigation, William Brown compiled a list of Newcomen engines known to have been erected in the North of England and Scotland. Smeaton chose a representative sample of fifteen of these engines and recorded their performance. A similar list was also compiled of the leading Cornish engines, this list is reproduced on page 64 of the current volume. The Cornish list records details of fifteen engines at work at Wheal Virgin, Poldice, Wheal Maid, Dolcoath and Wheal Chance in 1769.[4]

As a result of his experimental and empirical studies Smeaton was able to develop much more efficient and better engineered engines. Summarising Smeaton's findings Farey writes:

> "The principal cause of the defective performance of these engines was their imperfect execution and faulty proportions; the cylinders were very imperfectly bored, and therefore the pistons could not be fitted to move freely in them or made to fit tight, and consequently a leakage of water and air took place, as well as a great loss of power from friction; the pump was equally faulty.
>
> The boilers were too small to supply the cylinder properly with steam and ill – constructed; the fire - grates were placed too low beneath the bottom, which rose up very high in the middle; they had a most slovenly practice of heaping a great mass of coals on the fire – grate, making a most intense heat beneath the centre of the boiler, and not sufficient heat at the circumference; and very little attention was paid to keeping the boiler free from internal scales of crust. All these deficiencies tended to diminish the supply of steam, and caused the engines to move very slowly.
>
> The engineers who constructed these engines were very ignorant; and so far from seeing these defects, and removing them, they reckoned every thing to depend upon the size of the cylinder, and the load or burden per square inch of the piston, without taking into account the velocity of the motion, or making proper endeavours to gain the greatest mechanical power; all they looked to was to make engines of great force, which would lift great weights of water: hence the parts required to be very large, and strong, but they did not obtain an adequate result."[6]

Smeaton found that poor engineering was endemic, particularly when it came to accurately boring cylinders. Working with the Carron Company who had established a foundry near Falkirk in 1759 Smeaton was able to develop the necessary engineering

Table 2. Smeaton's list of atmospheric engines at work in Cornwall in 1769.

Names of mines	Horse power exerted	Diam. of cylinder Inches.	Area in square inches	Lbs. square inch	Motion per min. Feet	Millions lbs. per bushel	Bushels per month	Depth Drawn fathoms	Cubic feet drawn per minute
Wheal Virgin	18.80	60	2827	5.70	38.5	5.82	4670	35	47.35
Wheal Virgin	18.25	60	2827	5.54	38.5	5.30	4980	34	47.35
Wheal Virgin	22.08	60	2827	7.00	38.5	7.79	4284	43	47.35
Wheal Virgin	20.45	64	3217	7.63	27.5	7.47	3960	71	25.35
Wheal Virgin	29.65	70	3848	6.60	38.5	7.63	5616	43	60.80
Poldice	14.78	60	2827	5.23	33.0	7.17	2979	25	52.20
Poldice	19.51	66	3421	5.70	33.0	7.26	3945	33	52.20
Poldice	15.46	66	3421	5.42	27.5	7.81	2864	63	21.60
Wheal Maid	28.28	60	2827	6.66	49.5	6.76	6048	36	69.30
Wheal Maid	25.13	65	3318	5.05	49.5	7.42	4896	32	69.30
Dolcoath	17.88	60	2827	6.32	33.0			42	49.20
Dolcoath	16.65	63	3117	5.88	30.0			63	29.80
Dolcoath	20.93	70	3848	5.98	30.0			71	37.56
Wheal Chance	16.02	69½	3794	6.97	20.0	7.27	3186	51	28.00
Wheal Chance	16.66	70	3848	7.14	20.0	7.56	3186	56	28.00
Averages	**20.04**	**64½**	**3253**	**6.18**	**33.8**	**7.10**	**4210**	**46½**	**44.30**

Cited in Farey, J., 1971.

facilities to meet his exacting standards. This included a cylinder boring mill the design of which Smeaton submitted to the Carron Company in October 1770.[7] Farey comments that Smeaton's boring mills at Carron were "the most powerful machines at that time, and by which many large cylinders were bored for the atmospheric engines".[8] Whilst this mill later failed to satisfy James Watt's even more exacting standards, Smeaton's mill was producing much more accurately bored cylinders than those previously supplied by Coalbrookdale.

The first engine to employ Smeaton's newly derived principles was designed for Long Benton Colliery on Tyneside in 1772; this became the prototype for Smeaton's later engines. The Long Benton engine, which was started in February 1774, had a 52″ cylinder and a working stroke of seven feet and regularly made twelve stokes per minute. The engine worked two twenty-four fathom pump columns of twelve inch diameter. Writing of the Long Benton engine Farey notes:

> "This was the standard by which Mr Smeaton proportioned all his other engines. It performed extremely well; being a medium size, between the largest and smallest engines, its dimensions and proportions may be taken for a good example of Newcomen's invention in its most perfect state."[9]

No doubt the Cornish adventurers were watching Smeaton's activities with interest. As has been seen enlarging cylinder size or, as an alternative, shammeling were the typical Cornish solutions to the challenge of increasing depth. However these were very expensive solutions to the problem of draining deep mines. The great expense of keeping mines drained must have exercised the minds of Cornish adventurers, particularly in the light of the onslaught that Cornish copper mining was experiencing from Parys Mountain copper mines on Anglesey. In this context it is hardly surprising that John Smeaton was commissioned to design an engine in the county.

In 1774 – 1775 Smeaton designed a 72″ engine for the notoriously wet Chacewater Mine. Smeaton's 72″ was intended to replace two existing Newcomen engines recorded by Smeaton as being a 66″ and a 64″, both with a six foot stroke (Note: Farey[10] quotes 64″ and 62″). The 66″ worked twenty-four fathoms of 18½″ bucket pumps whilst the 64″ worked twenty six fathoms of 17½″ bucket pumps. The 66″ worked the lower column, shammeling to the 64″. The total lift being fifty fathoms, discharging to adit which was itself twenty four fathoms below surface. Between them they raised eighty cubic feet of water per minute in summer and one hundred cubic feet per minute in winter. The pair of engines were said to consume 16½ bushels of coal an hour.[11]

As would be expected the castings for the Chacewater engine were supplied by the Carron Company:

"*Particulars of the weights of some of the principal parts of the cast iron – work, for the Chase –water Fire – Engine; executed at Carron Iron – works, Scotland, 1775.*

The cylinder bored quite through, 72 inc. diam. 10½ ft. long, with a flange round its lower end, and a smaller flange round the top; also a very strong flange near the middle, to suspend it by, and fasten it to the beams. 4 tons, 16 cwt., 4 lbs.

The hemispherical cylinder – bottom, with its flange, to unite it to the cylinder; also a short steam – pipe, and necks for the injection – pipe, snifting – pipe, and eduction – pipe. 1 ton, 13 cwt., 36 lbs.

Cup, or circular channel, to fix the top of the cylinder. 13 cwt.

The cast – iron piston. 14 cwt., 10 lbs.

Cast – iron axis for the great lever. 16 cwt., 14 lbs.

Cast – iron part of the regulator, or steam valve. 1 cwt., 98 lbs.

Prices paid to the Carron Company for the above castings, delivered in Cornwall.

For the bored cylinder and working barrels, and for the regulator – valve, 28s per cwt.

For the clack and buckets, turned, 21s per cwt.

For the cylinder – bottom, piston, cup, clack door – pieces for the pumps, &c. 18s per cwt.

For the axis and similar castings, 16s per cwt.

For the common pipes for the pumps, and other plain pipe – work, and furnace doors, 14s. per cwt.

For grate bars, and bearer bars, and hearth plates, weights, &c. 11s per cwt.

For wrought – iron screw bolts and nuts, 5d per lb.

For brass castings for the pump clack seats, and the regulator – valve, injection – cock, &c. 16d per lb. without turning or fitting."[12]

As on the Long Benton engine the Chacewater engine's beam or "Great lever" was a complex structure of laminated timber, requiring precise workmanship:

"The great lever of this engine appears ..... to have been of prodigious strength. The depth in the centre was 74 inc. of which the iron axis occupied 5 inc. at the middle parts. The breadth was 24 inc. It consisted of twenty beams of fir, viz one above another .....and two such sets side by side. The depth of the lever at its extremities was 60 inc. by 24 inc. These twenty pieces were united by two large iron bands near the centre, 3½ inc wide, by ¾ inc. thick, with nuts at their upper ends to bind the wood fast; and also thirty – two iron bolts, 1 inc. diam., each of which passed through the whole depth of the beams ..... but there were four bolts abreast at each of those places. The upper and lower ends of these bolts, passed through sixteen strong iron plates, which were let into the wood at the top and bottom of the lever; and as each plate extended all across it breadth, they united the two sets of beams, which were side by side.

There were two large oak keys, 7 inc. by 5 inc. inserted between the four principal beams in the middle, and fifty four smaller keys, 2½ inc. by 5 inc. inserted between the joints of the different beams, towards the end of the lever ..... The cylindrical gudgeons or pivots, at the ends of the axis, on which the immense lever was poised, were 8½ inc diam. and 8½ inc. length of bearing; they rested in brass sockets, let into large blocks of wood, worked into the masonry of the lever wall; the length or distance between these bearings was 3½ feet."[13]

Such a substantial engine required a substantial house thirty six feet long and twenty feet wide externally and twenty three feet by fourteen and a half feet internally. The "great lever wall" was ten feet wide at its base, the upper part five feet. The house was entirely constructed of granite.[14]

Smeaton designed the Chacewater engine to work fifty one fathoms of 16¾ inch pumps at a rate of "full nine strokes a minute". According to his calculation this

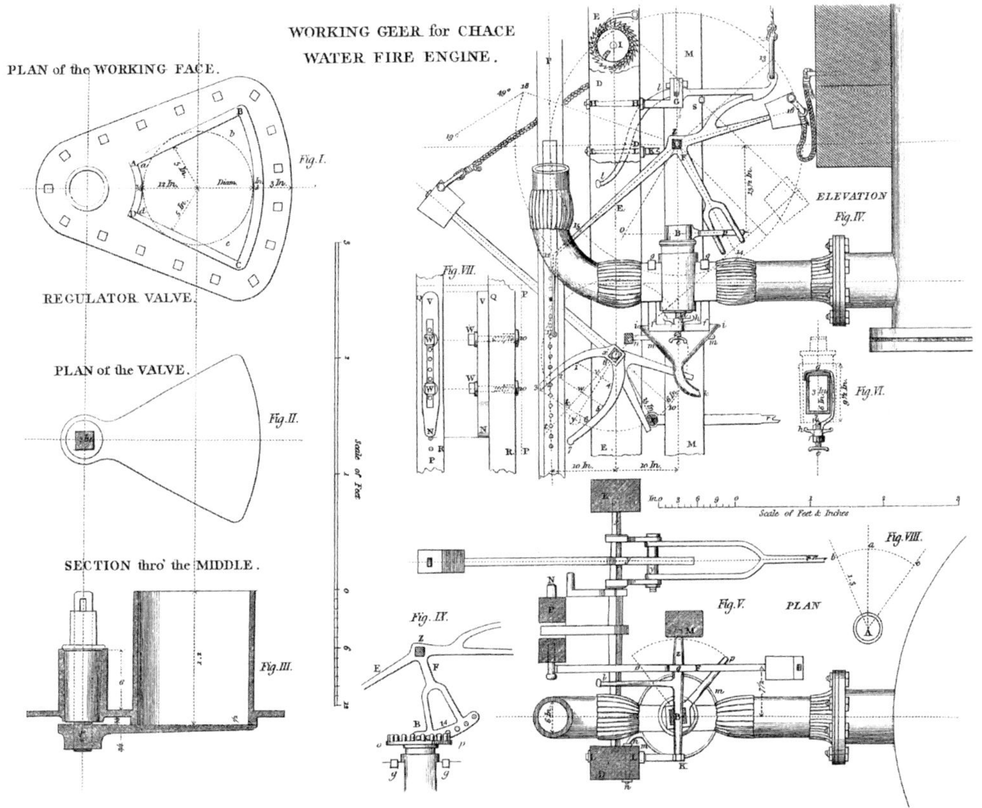

Details of the regulator and working "geer" of Smeaton's 72-inch at Chacewater.
Image courtesy of Tony Clarke.

would raise 880 hogsheads of water an hour. Assuming a hogshead of 52.5 imperial gallons this gives 123.6 cubic feet per minute. To achieve this output Smeaton anticipated that the engine would need thirteen bushels of coal (London measure) an hour. Smeaton notes that the boiler under the cylinder will not consume "above six bushels of coals per hour ...... with advantage". To make up the shortfall Smeaton proposed to use the heat "from the proposed smelting furnaces" to heat a second boiler which suggests that he must have been favourably influenced by Swaine's use of waste heat in conjunction with his moorstone boiler.[15] Farey comments:

> "...... and another boiler was made for Mr Smeaton's engine at Chase – water mine, to collect the waste heat of the furnaces; but it was made of iron – plates riveted together instead of stone, though the construction in other respects, was nearly the same as the stone boiler. The length was 22 feet, by 8½ feet wide, and 10¾ feet deep, with four internal tubes 22 inches diameter; two of them were fixed at twelve inches above the bottom, and the other two at 3¾ feet above the bottom. This boiler was intended to be used in addition to the centre boiler ....."[16]

This experiment does not appear to have been a success: William Pole, who got his information from a Mr Moyle* who had been employed on the mine at the time the engine was erected, observes:

> "It had, when originally erected, one boiler 15 feet diameter placed immediately under the cylinder, and an extra one, constructed to collect and make use of the waste heat from the furnaces upon the works. This latter however being found a failure, it was removed after a very short use, and two new boilers of the same construction and dimensions as the centre one were added, being fixed in low buildings on each side of the engine house. These were found successful in furnishing the engine with steam."[17]

As reconstructed the engine had three fifteen foot diameter boilers; the original boiler under the cylinder; the other two in low buildings, one on each side of the house.[18]

The engine was set to work in August 1776.[19] Teething problems aside, the Chacewater 72″ was a fine engine; it was certainly the largest engine erected in Cornwall during the eighteenth century and, arguably, represented the high water mark of Newcomen engine design. Samuel Smiles, writing in the 1860s, says of the engine that it was: "the finest and most powerful work of the kind which had until then been constructed....."[21] Unfortunately in practice the engine seems to have

***William Pole's informant may well have been Matthew Moyle of "Chasewater" who, so Thomas Wilson informs us, "was Agent of Chasewater Mine when wrought by two Engines of the old Construction".[20]**

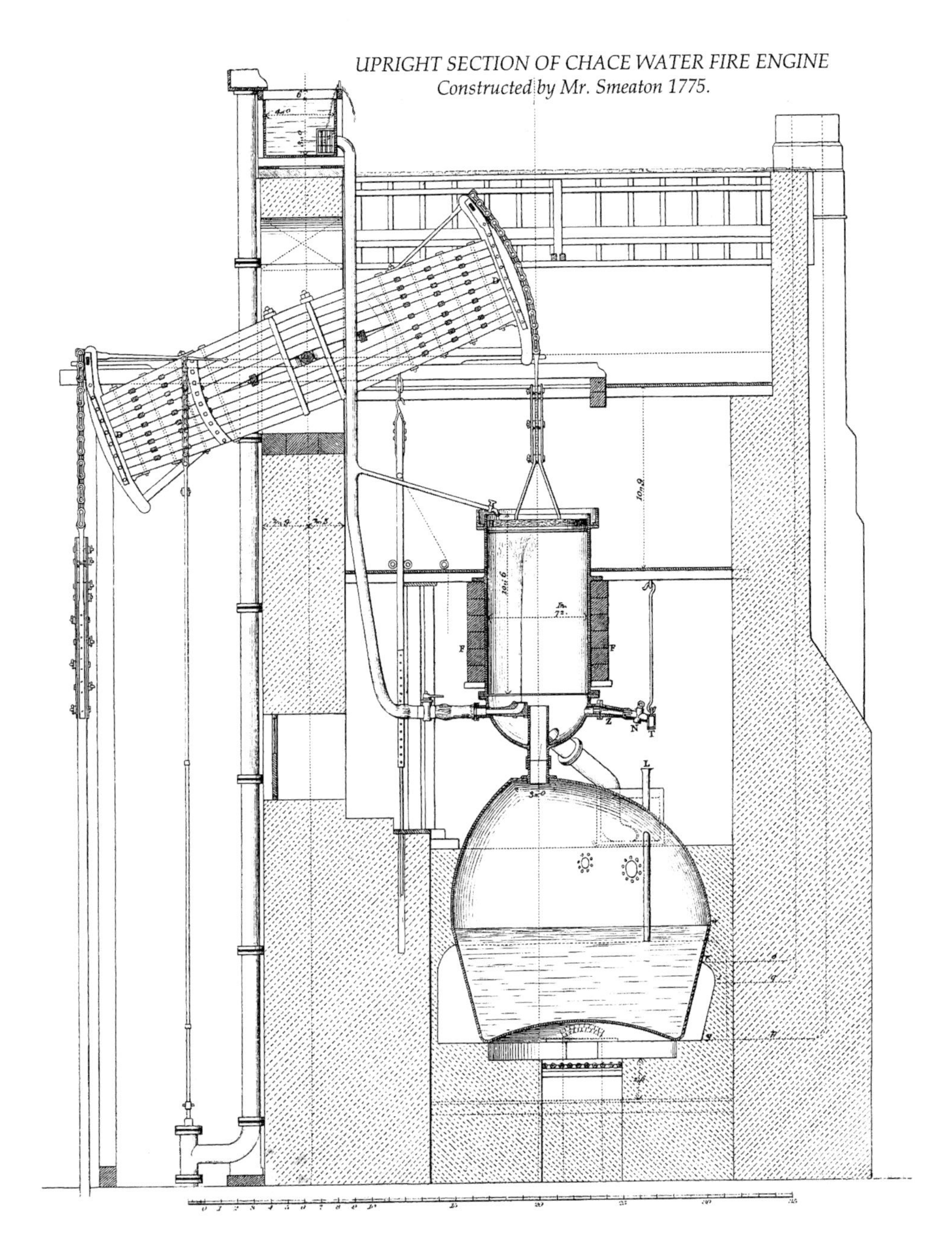

John Smeaton's 72-inch for the notoriously wet Chacewater Mine. Smeaton designed the engine in 1774-1775 and it was started in August 1776. In spite of being considered the high water mark in the development of the Newcomen engine the Chacewater 72-inch failed to bring the mine into fork and was soon replaced by a Boulton and Watt engine. ***Image courtesy of Tony Clark.***

failed to live up to its great promise. Thomas Wilson, who was latterly Boulton and Watt's agent in Cornwall, wrote in his undated (circa 1794) *Account of the state of the principal mines in the county of Cornwall at the time of the introduction of Mr Watts improved steam engines into the county*:

> "In 1773 Fenton & Co. took this mine & Mr Jno Smeaton, Engineer, having for some time before attended to the improvement of Steam Engines then in Use, they were induced to apply to him to build an Engine with his improvements, which Engine being the largest ever built in Cornwall was set to work in Aug(u) st. 1776, but after working several Months at considerable expense it was found ineffectual for the purpose wanted, having never been able to sink the water more than 2/3rds of the Debth required to work the Mine to effect, & the Engine was of necessity obliged to stop till a Level (*Author's note: County Adit*) which had been begun at near a mile distant from the Mine in 1773 could be finished, which Level was for a considerable distance 50 Fathoms under the Surface & which when compleated would take 14 Fathoms from the Depth of the Mine; Viz would reduce the height necessary to raise the water from 62 to 48 Fathom."[22]

Whilst the Chacewater engine must have been magnificent, Smeaton had reached or indeed exceeded the limits of Newcomen engine technology. Although Smeaton's engineering was exemplary he had not addressed the intrinsic flaw of the atmospheric engine: the massive thermal inefficiency inherent in the repeated heating and cooling of the cylinder. In Cornwall, where economic use of coal was fundamental, this would prove to be the final nail in the coffin for the atmospheric engine. The man who drove that nail would be James Watt and, to stretch the metaphor to breaking point, the hammer he used to drive the nail was the separate condenser which obviated the need to heat and cool the cylinder.

As something of a stop gap a Boulton and Watt engine was erected at Chacewater in 1777, the engine, a 30″, being started in September of that year. This was the first Boulton and Watt engine to be erected in Cornwall. The Chacewater 30″ appears to have favourably impressed Thomas Wilson, who at that time was employed at Chacewater by Fenton and Co. Wilson advised his employers that rather than waiting three years for the County Adit to reach the mine they should erect a larger Boulton and Watt engine which "would do all the business that was intended to have been done by the Engine before built by Mr Smeaton".[23] Fenton and Co. followed this advice and in August 1778 Smeaton's 72″ was rebuilt by Boulton and Watt as a 63″, the original cylinder serving as a steam jacket.[24] Wilson notes:

> "......& tho it was certainly not so perfect as if it had been entirely new, yet it got out all the water to the Depth of 62 Fathoms, and also enable the Co(mpany) to sink to the Depth of 72 Fathoms before the level was completed, which was not done till Jan(uary) 1780."

By way of a postscript to John Smeaton's activities in Cornwall, the Chacewater 72″ was not the last of these engines built. At least three other 72″ engines to Smeaton's design were erected after the Chacewater engine which suggests that the design was sound. These include: a 72″ "returning engine" for the Carron Ironworks, and 72″ pumping engines for Middleton Colliery in Yorkshire and Bourn Moor Colliery in the north east.[25] In summing up Smeaton's contribution to the evolution of the Newcomen engine Pole concludes:

> "He does not seem to have added anything new to the machine, or to have invented any thing connected with it which can be easily particularised; but by a careful study of its action, and an accurate theoretical consideration of the several relations of its parts one to the other, he so contrived to improve their proportions as to increase the duty by this means alone nearly fifty per cent.[26]

Chapter 8 References

1. See Smeaton J., 1759
2. Cited in Farey J. 1827
3. Farey J., 1827
4. Allen & Rolt, 1997; Farey J., 1827; Farey J, 1971
5. Cited in Farey J., 1971
6. Farey J. 1827
7. Smeaton J., 1812a
8. Farey J., 1827
9. *Ibid*
10. *Ibid*
11. *Ibid*; Smeaton J., 1812b
12. *Ibid*
13. *Ibid*
14. *Ibid*
15. Smeaton J., 1812b
16. Farey J., 1827
17. Pole W., 1844
18. Farey J., 1827
19. Wilson papers: AD1583/11/68
20. Wilson T., 1792
21. Smiles S., 1865
22. Wilson papers: AD1583/11/68
23. *Ibid*
24. *Ibid*; Dickinson H. W. & Rhys Jenkin, 1927
25. Farey J., 1827; Skempton A. W., 1981
26. Pole W., 1844

Chapter 9
Newcomen engines erected in Cornwall

This list should not be considered definitive, further work is needed to clarify ambiguities. It is probable that some engines on this list appear more than once in slightly different guises. This is particularly the case in mines with multiple engines such as Dolcoath and North Downs and further work is required to establish more robust details of engines that worked there. Also further research into customs records (for example) will undoubtedly yield details of further engines. The strength of this list is that it demonstrates just how widespread the Newcomen engine was in Cornwall in the middle years of the eighteenth century, demonstrating that Pryce's estimate of three score engines may well have been on the conservative side.

Table 3. List of Newcomen engines erected in Cornwall.

Mine	Date	Cylinder	Notes	References
1. Ale & Cakes	1763/4 & 1767		Fire engine castings supplied by Coalbrookdale	Rogers K. M., 1976
2. Balcoath	Pre 1712		Vague tradition of very early engine; mentioned in Cunnack mss	Hamilton Jenkin, 1927; Brooke J., 1993
3. Bosprowal	By 1758		One of the engines mentioned by Borlase	Borlase W., 1758
4. Bullen Garden	By 1758		One of the engines mentioned by Borlase; may be one of the engines recorded as being at Dolcoath	Borlase W., 1758; Pryce W., 1778
5. Bullen Garden	By 1778		Second engine shown of Pryce's Bullen Garden section	Pryce W., 1778
6. Carloose		45″	Re-erected at Dolcoath 1775	Trevithick F., 1872.
7. Chacewater	1725 – 7		Erected by Joseph Hornblower	Allen & Rolt 1997; Redding C., 1833
8. Chacewater	1750	54″	Cylinder supplied by Coalbrookdale	Berg T. & P., 2001; Rogers K. M., 1976

9. Chacewater	1766	66″	Cylinder supplied by Coalbrookdale; offered for sale July 1770 along with 64″ (see 8 below)	Barton D. B., 1966 citing *Sherborne Mercury* 23.7.1770; Rogers K. M., 1976
10. Chacewater	By 1770	64″	July 1770 offered for sale along with 66″	Barton D. B. citing *Sherborne Mercury* 23.7.1770
11. Chacewater	By 1770	62″	Pumping from same shaft as 64″	Farey J., 1827
12. Chacewater	1775	72″	Smeaton's 72″	Farey J., 1827
13. Wheal Chance	By 1777	70″	One of Smeaton's 15 Cornish engines	Farey J, 1971; Wilson T., 1792
14. Wheal Chance	By 1777	69½″	One of Smeaton's 15 Cornish engines	Farey J, 1971; Wilson T., 1792
Consolidated (see Wheal Maid & Wheal Virgin)			Seven engines	
15. Danniell & Co. (unspecified mine)	1761		Fire engine castings supplied by Coalbrookdale to Thos Danniell & Co.	Rogers K. M., 1976
16. Dolcoath	1746	40″	Possibly one of the engines mentioned by Borlase; this may be the Bullen Garden engine (see 4 above); cylinder supplied by Coalbrookdale	Borlase W., 1758; Rogers K. M., 1976
17. Dolcoath	1753	54″	Possibly one of the engines mentioned by Borlase; this may be the Bullen Garden engine (see 4 above); cylinder supplied by Coalbrookdale	Borlase W., 1758; Rogers K. M., 1976
18. Dolcoath	1768	63″	Cylinder supplied by Coalbrookdale; one of Smeaton's 15 Cornish engines	Farey J, 1971; Rogers K. M., 1976
19. Dolcoath	1771	70″	Erected by John Wise and John Budge; possibly ex Wheal Weeth (see below); one of Smeaton's 15 Cornish engines.	Buckley A., 2010; Farey J., 1971
20. Dolcoath		60″	One of Smeaton's 15 Cornish engines.	Farey J., 1971
21. Dolcoath	1775	45″	"Dolcoath New Engine" erected by John Budge; ex Carloose.	Trevithick F., 1872.
22. Drannack	1748	55″	One of the engines mentioned by Borlase	Rogers K. M., 1976

23. Drannack	By 1767	63″	Sept. 1767 offered for sale	Barton D. B., 1966 citing *Sherborne Mercury* 14.9.1767
24. Wheal Fortune (Ludgvan)	1720	47″	See main text; nineteenth century writers have confused this with the 47″ Ludgvan Lez/Lease engine	For a well argued discussion of this engine see Brooke J., 1996
25. Wheal Fortune (Redruth)	By 1770	70″	Dec 1770, two 70″ engines, one lately erected; April 1771 both 70″ engines offered for sale	Barton D. B., 1966 citing *Sherborne Mercury* 31.12.1770 & 8.4.1771
26. Wheal Fortune (Redruth)	By 1770	70″	Dec 1770, two 70″ engines, one lately erected; April 1771 both 70″ engines offered for sale	Barton D. B., 1966 citing *Sherborne Mercury* 31.12.1770 & 8.4.1771
27. Wheal Fortune (St. Hilary)	1774/5	65″	Offered for sale. June 1778	Barton D. B., 1966 citing *Sherborne Mercury* 29.6.1778
28. Godolphin			Possibly the Huel Rith engine?	Wilson T, 1792
29. Great Work	1754	47″	Cylinder ex Ludgvan Lease; erected John Nancarrow	Rogers K. M., 1976
30. Great Work	By 1773	63″	Offered for sale September 1773	Barton D. B., 1966 citing *Sherborne Mercury* 9.6.1773
31. Herland	1753	70″	One of the engines mentioned by Borlase; offered for sale Sept 1767	Barton D. B., 1966 citing *Sherborne Mercury* 14.9.1767; Borlase W., 1758; Rogers K. M., 1976.
32. Herland		60″	Offered for sale Sept 1767	Barton D. B., 1966 citing *Sherborne Mercury* 14.9.1767
33. Hewas	By 1771	52″	Offered for sale Sept. 1771	Barton D. B., 1966 citing *Sherborne Mercury* 23.9.1771 & 19.9.1774
34. Hewas	By 1771	55″	Offered for sale Sept. 1771	Barton D. B., 1966 citing *Sherborne Mercury* 23.9.1771 & 19.9.1774

35. Higher Rose-warne & Wheal Gerry	1768		Fire engine castings supplied by Coalbrookdale	Rogers K. M., 1976
36. Killycor & Poldice (see 51-53 below)	Circa 1765		August 1765 "upon which one fire engine has lately been erected and one more is now building"	Barton D. B., 1966 citing *Sherborne Mercury* 31.8.1765
37. Killycor & Pol-dice (see below)	Circa 1765		August 1765 "upon which one fire engine has lately been erected and one more is now building"	Barton D. B., 1966 citing *Sherborne Mercury* 31.8.1765
38. Lemon & Co. (unspecified mine)	1746	40″ (esti-mated)	Fire engine castings supplied by Coalbrookdale to Lemon & Co.	Rogers K. M., 1976
39. Lemon & Co. (unspecified mine, possibly Polgooth)	1749	52″	Fire engine castings supplied by Coalbrookdale to Lemon & Co.	Rogers K. M., 1976
40. Ludgvan Lease	1746	47″	Cylinder supplied by Coal-brookdale; one of the engines mentioned by Borlase; moved to Great Work 1754	Borlase W., 1758; Rogers K. M., 1976
41. Wheal Maid		60″	One of Smeaton's 15 Cornish engines	Farey J., 1971; Wilson T., 1792
42. Wheal Maid		65″	One of Smeaton's 15 Cornish engines	Farey J., 1971; Wilson T., 1792
43. Metal Work (Chacewater)	1750 (?)	54″ (?)	One of the engines mentioned by Borlase; probably the Chacewater 50″ (see 8 below)	Borlase W., 1758; Rogers K. M., 1976
44. North Downs	By 1754		"Old engine" seen by R. R. Angerstein in spring 1754.	Berg T. & P., 2001
45. North Downs	By 1754	69″	Seen by R. R. Angerstein, spring 1754; this is almost certainly, see 67 below.	Berg T. & P., 2001
46. North Downs	1756	60″	Cylinder supplied by Coal-brookdale; Borlase mentions two engines on North Downs in 1758	Borlase W., 1758; Rogers K. M., 1976; Berg T. & P., 2001
47. North Downs	By 1758		Borlase mentions two engines on North Downs in 1758	Borlase W., 1758
48. North Downs	By Sept 1760		Sept 1760: five fire engines on North Downs	Barton D. B., 1966 citing *Sherborne Mercury* 8.9.1760
49. Wheal Oula (North Downs)	1763	70″	Cylinder supplied by Coal-brookdale	Rogers K. M., 1976

50. Owen Vean (Marazion)	1762		Fire engine castings supplied by Coalbrookdale	Rogers K. M., 1976
51. Wheal Park (Marazion)	Circa 1771	67″	Fire engine lately erected in April 1771; disposed in lots May 1772	Barton D. B., 1966 citing *Sherborne Mercury* 8.4.1771
52. Pednandrea	1760-61		Fire engine castings supplied by Coalbrookdale	Rogers K. M., 1976
53. Pittslooarn (Chacewater)	By 1758		One of the engines mentioned by Borlase; possibly Hornblower's Chacewater engine of 1725-7.	Borlase W., 1758
54. Poldice	1763	60″	Cylinder supplied by Coalbrookdale; one of Smeaton's 15 Cornish engines; 1778 Tested by Boulton & Watt.	Farey J., 1971; Rogers K. M., 1976; Wilson T., 1792
55. Poldice		66″	One of Smeaton's 15 Cornish Engines; 1778 one of the two 66″ engines tested by Boulton & Watt.	Farey J., 1971; Rogers K. M., 1976; Wilson T., 1792
56. Poldice		66″	One of Smeaton's 15 Cornish engines; 1778 one of the two 66″ engines tested by Boulton & Watt.	Farey J., 1971; Rogers K. M., 1976; Wilson T., 1792
57. Poldory	1761-62 & 1764		Fire engine castings supplied by Coalbrookdale	Rogers K. M., 1976
58. Polgooth	1725-27		Erected by Joseph Hornblower	Allen & Rolt 1997; Redding C., 1833
59. Polgooth	By 1754	40″ (?)	Two engines noted by R. R. Angerstein in spring 1754; mentioned in Borlase (See 40 above)	Berg T. & P., 2001
60. Polgooth	1747	40″	Cylinder supplied by Coalbrookdale; mentioned in Borlase	Borlase W., 1758; Rogers K. M., 1976
61. Pool Adit	1748	60″	Cylinder supplied by Coalbrookdale; one of the engines mentioned by Borlase.	Berg T. & P., 2001; Borlase W., 1758; Rogers K. M., 1976
62. Wheal Reeth in Godolphin Bal	By 1758		One of the engines mentioned by Borlase; possibly the Godolphin engine.	Borlase W., 1758
63. Wheal Rose			One of the engines mentioned by Borlase (Is this the 1753 60″ ? see 67 below)	Borlase W., 1758

64. Wheal Rose (North Downs)	1725		Erected by Joseph Hornblower	Allen & Rolt 1997; Redding C., 1833
65. Wheal Rose	1753	60″	Cylinder supplied by Coalbrookdale (is this the Huel Ros engine? See 65 above); probably Angerstein's 69″ seen in 1754 at North Downs see 46 below	Rogers K. M., 1976
66. Roskear	1746	47″	Cylinder supplied by Coalbrookdale	Rogers K. M., 1976
67. Wheal Sparnon	1764		Fire engine castings supplied by Coalbrookdale; 1765 engine being erected by Hornblower	Harris T. R., 1976; Rogers K. M., 1976.
68. Stowe's Mine	Circa 1730		Probably a figment of D. B. Barton's imagination	Barton D. B., 1964; Messenger M., 2015
69. Treleigh Wood (Redruth)	1773	47½″	Cylinder supplied by Coalbrookdale	Rogers K. M., 1976
70. Tresavean	1765	60″	Cylinder supplied by Coalbrookdale	Rogers K. M., 1976
71a. Trevenson	1744	40″ (estimated)	Flawed cylinder supplied by Coalbrookdale	Rogers K. M., 1976
71b. Trevenson	1745	40″ (estimated)	Replacement cylinder for 74a supplied by Coalbrookdale	Rogers K. M., 1976
72. Truan Tin Mine	By 1754		Noted by R. R. Angerstein, spring 1754	Berg T. & P., 2001
73. Wheal Virgin	1758	60″	Cylinder supplied by Coalbrookdale; one of Smeaton's 15 Cornish engines; known as "Old" engine	Farey J., 1971; Rogers K. M., 1976
74. Wheal Virgin	1765	60″	Cylinder supplied by Coalbrookdale; one of Smeaton's 15 Cornish engines	Farey J., 1971; Rogers K. M., 1976
75. Wheal Virgin		60″	One of Smeaton's 15 Cornish engines	Farey J., 1971
76. Wheal Virgin		64″	One of Smeaton's 15 Cornish engines; known as "West" engine	Farey J., 1971
77. Wheal Virgin		70″	One of Smeaton's 15 Cornish engines; known as "New" engine	Farey J., 1971
78. Wheal Weeth			Sampson Swaine – Moorstone boiler	Carter C., 1986; Harris T. R., 1975

Chapter 10
Boulton and Watt: The 1769 patent.

It has been noted that the inherent flaw of Newcomen's atmospheric engine was the thermal inefficiency resulting from the repeated heating and cooling of the cylinder. In 1763 James Watt (1736 – 1819), then a Scottish mathematical instrument maker, was engaged to repair a small model of a Newcomen engine on behalf of Glasgow University. Watt was struck by how inefficient the engine was and set about through a series of experiments to determine why. He concluded that to make the best use of steam, the temperature of the cylinder needed to be maintained at the temperature of the steam. This was totally counter to the working principle of the Newcomen engine which required the rapid cooling of the cylinder during the condensing phase of the engine's operating cycle. Watt's solution to this conundrum was sublime. In 1765 Watt developed the idea of condensing the steam in a separate vessel to the cylinder; the separate condenser was born.[1] In a letter written to James Watt on 5th February 1778 no less a luminary than John Smeaton commented:

> "Your idea of condensing in a separate vessel from the cylinder...... I look upon as a greater Stroke of invention, than has appeared since Newcomen."[2]

James Watt obtained his first patent, No. 913, on 5th January 1769 for a period of 14 years. The specification is outlined below in some detail because it is such an important element in the story of the pumping engine in Cornwall in the latter years of the eighteenth century:

> "To all to whom these presents come, &c. Now know ye, that my lessening the consumption of steam, and consequently fuel, in fire engines, consists of the following principles:
>
> First, that vessel in which the powers of steam are to be employed, to work the engine which is called *the cylinder* in common fire engines, and which I call the *steam vessel*, must, during the whole time the engine is a work, be kept as hot as the steam that enters it; first by enclosing it in a case of wood, or any other materials that transmit heat slowly; secondly, by surrounding it with steam, or other heated bodies; and thirdly by suffering neither water, nor any other substance colder than the steam, to enter or touch during that time.

James Watt (1736 – 1819).

Secondly, in engines that are to be worked wholly or partially by condensation of steam, the steam is to be condensed in vessels distinct from the steam vessels, or cylinders, although occasionally communicating with them; these vessels I call *condensers*; and, whilst the engines are working, these condensers ought at least to be kept as cold as the air in the neighbourhood of the engines, by application of water and other cold bodies.

Thirdly, whatever air or other elastic vapour is not condensed by the cold of the condenser, and may impede the working of the engine, is to be drawn out of the steam – vessels, or condensers by means of pumps, wrought by the engines themselves, or otherwise.

Fourthly, I intend, in many cases, to employ the expansive force of steam to press on the pistons, or whatever may be used instead of them, in the same manner as the pressure of the atmosphere is now employed in common fire – engines: in cases where cold water cannot be had in plenty, the engines may be wrought by this force of steam only, by discharging the steam into the open air after it has done its office.

Fifthly where motions round an axis are required, I make the steam vessels in the form of hollow rings, or circular channels, with proper inlets and outlets for the steam, mounted on horizontal axles, like the wheels of a water mill; within them are placed a number of valves, the suffer any body to go round the channel in one direction only; in these steam – vessels are place weights, so fitted to them as entirely to fill up a part or portion of their channels, yet rendered capable of moving freely in them, by the means herein – after mentioned or specified. When the steam is admitted in these engines, between these weights and the valves, it acts equally on both, so as to raise the weight to one side of the wheel, and, by the re – action on the valves, successively, to give a circular motion on the wheel, the valves opening in the direction of which the weights are pressed, but not in the contrary; as the steam – vessel moves round, it is supplied with steam from the boiler, and that which has performed its office may either be discharged by means of condensers, or to the open air.

Sixthly, I Intend, in some cases, to apply a degree of cold, not capable of reducing the steam to water, but of contracting it considerably, so that the engines shall be worked by the alternate expansion and contraction of the steam.

Lastly, instead of using water to render the piston or other parts of the engine air and steam tight, I employ oils, wax, rosinous bodies, fat of animals, quicksilver, and other metals, in their fluid state.[3]

The 1769 specification is somewhat vague and this vagueness would have significant implications for Boulton and Watt and for Cornish mine adventurers and engineers particularly during the 1790s. The vagueness of the patent was, counter-intuitively, one of its great strengths; because it was so vague it was in effect almost all encompassing and consequently very difficult to circumvent. That the specification was vague is not surprising given that Watt had yet to try his principles on anything but an experimental level. At the date the patent was granted, Watt had not managed to prove his principles on a full size working engine. From 1767 Dr. Roebuck had provided Watt with the financial support to erect a full size engine at

Matthew Boulton (1728 – 1809). Boulton's business acumen allowed James Watt to translate his ideas into a very profitable reality.

Kinneil, however progress was slow due to the necessity for Watt to earn his living as a civil engineer. By 1769 the engine had been erected but it could not be made to work in a satisfactory manner. The problems were compounded by Roebuck's increasingly precarious financial position, culminating in his bankruptcy.[4]

Roebuck's problem was Matthew Boulton's opportunity; Boulton (1728 – 1809)

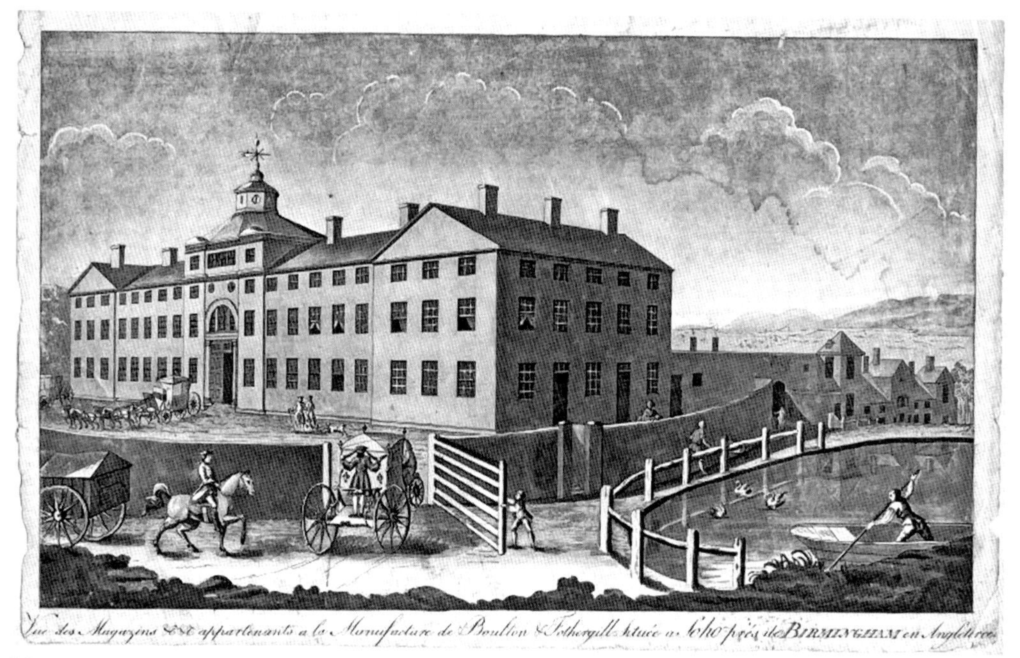

Matthew Boulton's Soho Manufactory.

acquiring Roebuck's interests in Watt's invention in 1773. For Watt this was the best possible outcome, Boulton was a successful Birmingham manufacturer with the resources to develop Watt's engine. Matthew Boulton's Soho Manufactory was described as "already the most complete manufactory for metal work in England, and conducted with the most spirit".[5] The Kinneil engine was transferred to the Soho Manufactory and by 1774 it was working as intended. However by the time the Kinneil engine was up and running at Soho five years of the original fourteen year patent had already expired. Thus on 23rd February 1775 Watt presented a Bill to the House of Commons to extend the original 1769 patent for a further period of twenty five years. The Bill received Royal assent on 22nd May 1775, extending Watt's patent to 1800 and ensuring Boulton and Watt's near monopoly on engine building in Cornwall and beyond. Having been granted the extension, Boulton and Watt formally entered into a partnership on 1st June 1775.[6]

It is worth noting that Boulton and Watt were not suppliers of complete engines. Typically they would produce drawings and the Soho Manufactory would produce the smaller parts including valves, nozzles and pistons. Larger parts including cylinders were outsourced to iron masters, most notably John Wilkinson who had developed an improved cylinder boring mill in about 1775. Miscellaneous ironwork, woodwork and masonry was the province of local craftsmen. The erection of the engine would typically be overseen by one of Boulton and Watt's trained erectors. In the case of Cornwall Boulton and Watt sent men such as William Murdock, Tom Pearson and Edward Bull to the county to erect their engines. All the costs of the engine would be met by the customer. Whilst Boulton and Watt would derive some income from the design of engines and the supply of parts the most important element of their income came from the premiums payable by users of their engines for the right to use the technology covered by the 1769 patent. It was these premiums that would become so contentious in Cornwall.[7]

By June 1775 Boulton and Watt had secured orders to design two large engines: The first was a blowing engine with a 38″ cylinder for John Wilkinson's blast furnace at New Willey near Broseley. The second, and from a Cornish perspective the more important of the two, was a 50″ pumping engine for Bloomfield Colliery near Tipton. That engine was started on 8th March 1776, the event being recorded in *Aries's Birmingham Gazette* of 11th March 1776:

> "On Friday last a Steam Engine constructed upon Mr. Watt's new Principles was set to work at Bloomfield Colliery, near Dudley, in the Presence of its Proprietors, Messrs Bently, Banner, Wallin, and Westly; and a Number of Scientific Gentlemen whose Curiosity was excited to see the first Movements of so singular and so powerful a Machine; and whose Expectations were fully gratified by the Excellence of its performance. The Workmanship of the Whole did not pass unnoticed, nor unadmired. All the Iron Foundry Parts (which were unparalleled for truth were executed by Mr Wilkinson; the Condenser, with the Valves, Pistons and all the small Work at Soho by Mr Harrison and others; and the Whole was erected by Mr. Perrins, conformable to the Plans and under the direction of Mr. Watt. From the first Moment of its setting to Work, it made about 14 to 15 Strokes per Minute, and emptied the Engine Pit (which is about 90 Feet deep, and stood 57 Feet high in Water) in less than an hour."[8]

The Bloomfield engine was the first large pumping engine erected to Boulton and Watt's designs. Dickinson and Jenkin make the point that "the engine was undoubtedly meant to serve as a pattern to convince mine adventurers elsewhere, *e.g.* in Cornwall, of its value".[9]

Chapter 10 References

1. Dickinson H. W. & Jenkin R., 1927; Farey J., 1827
2. Cited in Skempton A. W., 1981
3. Cited in Anon 1794
4. Dickinson H. W. & Jenkin R., 1927
5. Farey J., 1827
6. Dickinson H. W. & Jenkin R., 1927
7. *Ibid*; Farey J., 1827
8. Cited in Dickinson H. W., & Jenkins. R. 1927
9. Dickinson H. W. & Jenkin R., 1927

Section of eighteenth century, forged, wrought iron chain preserved at King Edward Mine. Chain of this type was universally used on engines throughout the eighteenth century, connecting the pitwork, piston and plug rod to the main beam. ***Photo courtesy Pete Joseph.***

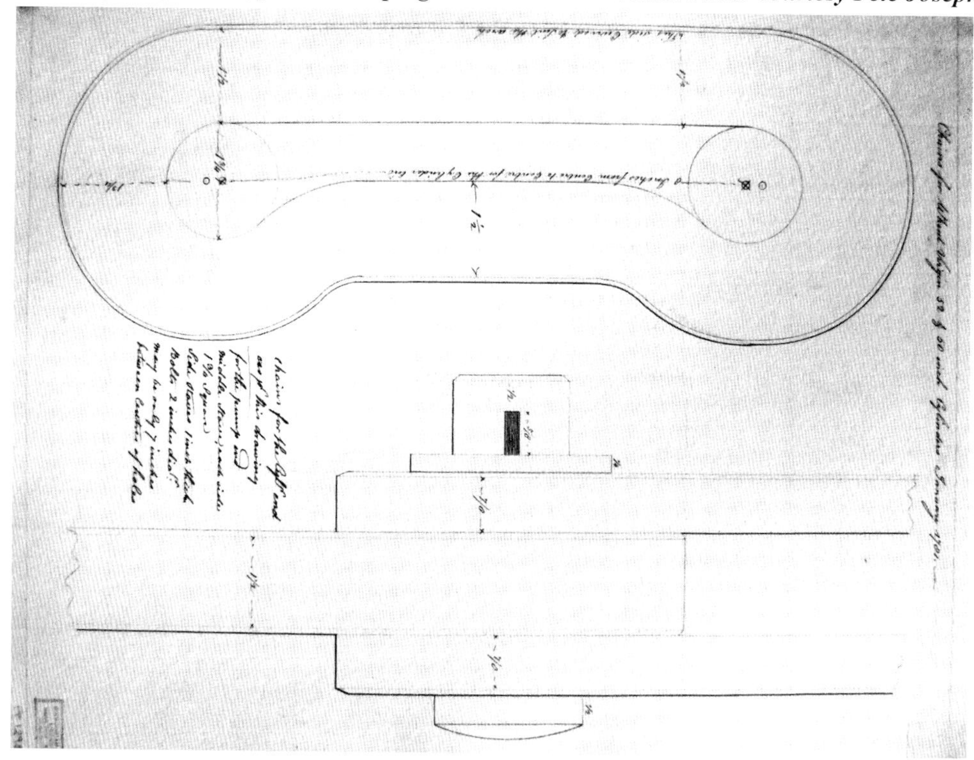

Boulton and Watt drawing of chain link for Wheal Virgin. Reproduced with the permission of the Library of Birmingham, MS 3147/1298.

Chapter 11
Boulton & Watt arrive in Cornwall.

By the mid 1770s it was increasingly obvious to Cornish adventurers that, even with expedients such as shammeling and increases in cylinder size, the limits of the Newcomen engine had been reached. The failure of Smeaton's 72″ to bring Chacewater / Wheal Busy into fork had only underlined the problem. William Pryce was writing at this precise time and was uniquely positioned as an observer. Like most of those involved in Cornish mining Pryce, as a mine adventurer, was painfully aware of the limitations of the Newcomen engine:

> "...... The vast consumption of fuel in those engines, is an immense drawback upon the profits of our Mines. It is a known fact, that every fire engine of magnitude consumes to the amount of three thousand pounds worth of coal in every year. This heavy tax on Mining, in some respects, amounts to a prohibition. No wonder then, that we should be more desirous to lessen the expense of maintenance in those devouring automatons, than frugal in their erection. Many trials of mechanical skill have been made by our engineers, to very little purpose, for the total application of heat and the saving of fuel. The fire place has been diminished, and enlarged again; the flame has been carried round the bottom of the boiler in a spiral direction and conveyed through the body in a tube (one, two or three) before its arrival to the chimney; some have used a double boiler, so that fire might act in every possible point of contact; and some have built a Moorstone boiler, heated by three tubes of flame passing through it."[1]

Fortunately for the Cornish mining interest a solution was at hand; Pryce observes:

> "Indeed the only improvement which has been made in the fire engine for thirty five years past, the publick will very justly attribute to the sagacity of Mr Watt, whose skill in pneumaticks, mechanicks, and hydraulicks, is evinced by the powerful application of elastick vapour, and by making a more perfect vacuum, nearly like that of a barometer, in his new constructed fire engine."[2]

James Watt's contribution to the development of the steam engine is too well known to linger over here. Of particular interest to the Cornish was the fact that

Watt had successfully addressed the massive thermal inefficiency of the Newcomen engine. Key to this was his invention of the separate condenser. This allowed steam to be condensed outside the cylinder obviating the need to heat and then cool the cylinder at every cycle as was the case with the Newcomen engine. Watt also introduced steam jacketing of the cylinder to retain heat. In practical terms this meant a considerable saving in coal. Pryce comments:

> "The working of these engines is more regular and steady than the common ones (*i.e. Newcomen engines*), and from what has been said, their other advantages are apparently very considerable; but to say exactly how much they excel common engines, is difficult as common engines differ very much among themselves. I am told, that the savings amount to least two – thirds of the fuel, which is a very considerable object where coals are as expensive as they are in Cornwall."[3]

Cornish mine adventurers were not slow to realise the potential of Watt's new engines. In the middle of 1776 a group of Cornish mining men, led by one Thomas Ennis, went to Birmingham to discover all they could about these engines. No doubt the newly started Bloomfield Colliery engine would have been on their itinerary. After the party had visited Matthew Boulton's Soho foundry it was discovered that an engine drawing was missing; it later transpired that Richard Trevithick Senior was the guilty party. Not surprisingly Boulton was less than impressed, writing to Ennis: "we do not keep a school to teach fire – engine making, but profess the making of them ourselves".[4]

The deputation must have reported favourably on its return to Cornwall. First off the mark was Jonathan Hornblower (I) who wrote to Matthew Boulton on 10th October 1776 on behalf of the Ting Tang adventurers:

> "Having heard of your improved Fire Engine I send you this to inform you that there is an engine to be erected next spring of the year in this neighbourhood the cylinder 52 inches diam: and that the Advents concerned have requested me (who am their Engineer & have a small share in the adventure) to desire you to send them proposals."[5]

Evidently Boulton and Watt's proposals must have been satisfactory for on 31st November 1776 Matthew Boulton wrote to James Watt:

> "We have a positive order for an engine for Ting Tang Mine, and from what I heard this day from Mr Glover, we may soon expect other orders from Cornwall. Our plot begins to thicken apace, and if Mr Wilkinson don't bustle a little as well as ourselves, we shall not gather our harvest before sunset."[6]

A significant amount of contemporary documentation regarding the Ting Tang engine survives. Of particular interest is *a list of all parts of the engine distinguishing where they are to be made*:

> "Bersham. Inside and outside cylinders, bottoms, piston, lid and stuffing box. The condenser cast iron work complete.
>
> Bradley. The nozzles and short pipe. The gudgeon and bed, the plummer block and brasses. Cast iron chains for cylr. end.
>
> Soho. The regulators, brass and ironwork. The F & Y, the plug frame chains, two chains for condenser pumps, gearing the buckets and clacks of the condenser, two round piston rods for do., all the screws and burrs for the two cylinders, condenser and boiler steam pipes; screwed ends for holding down screws, the iron work for the great piston rod top, the piston rod itself, 3 two inch diam. adjusting screws and burrs for top of piston chains, 3 do. for pump chains, if you chuse them. 4 screwed ends for the stirrup to hang the beam in, 2 ½ screws, screwed ends to shoot to martingale bolts.
>
> Tingtang. The boiler, firedoors and grates. The pumps and ye appurtenances. The pithead iron work, the poise beam, the pump chains, the screw bolts and arch plates for beam, the catch pins, the two glands which keep down the gudgeon, the stirrups for do., excepting the screwed ends which will be brought ready to shoot to. The martingales. The house and all the wood work.[7]

Both Bersham and Bradley foundries were owned by John Wilkinson. Although the Ting Tang engine was the first Boulton and Watt engine ordered by a Cornish mine, it was not the first put to work in Cornwall. As has already been noted, that honour goes to the 30″ engine erected on Wheal Busy / Chacewater. The castings, including cylinders for both these engines, were ready in May 1777 at John Wilkinson's Bersham Ironworks. The 52″ cylinder for the Ting Tang engine, was found to be too large for the hatches of the available vessel so the smaller Wheal Busy engine was shipped first. The delay in despatching the Ting Tang cylinder meant that this engine does not appear to have been put to work until December 1779. In contrast to the Ting Tang engine the Wheal Busy engine was "in considerable forwardness" when James Watt arrived to inspect it in August 1777, the engine being at work by mid September 1777 when Watt reported that "it has forked the water in the engine shaft".[8] A new era had begun.

In addition to meeting the cost of erecting Boulton and Watt's new engines, Cornish adventurers were also required to pay to Boulton and Watt premiums amounting to one third of the savings in fuel achieved by the new engine over a Newcomen engine. To establish a baseline, extensive tests were undertaken during the summer of 1778 by Boulton and Watt and the representatives of various Cornish mines. The trials were conducted on two engines at Poldice which were reputed to be the best in Cornwall:

> "Poldice Mines, October 30, 1778.
>
> We the under subscribers having carefully examined the books of the mine touching the consumption of coals of the two eastern engines during the months of August and September of this present year, and attended to the working of the engines during those months, do hereby certify, that the said two eastern engines did consume in 61 days of those two months 220 weys of coal, each wey being 64 Winchester coal bushels, which amount to 14,080 bushels in 61 days.
>
> We do also certify, that the said two engines together do work pumps in four lifts which are 17 inches diameter in the working barrels, and the whole depth from whence these pumps drew water to the adit is 58 fathoms.
>
> We further certify, that the said engines did during those months of August and September last, work the said pumps at the rate of 6 strokes of 5½ feet long each in every minute, which amounts to 8640 strokes per 24 hours.
>
> We have also made an accurate calculation, by which it appears that when the new fire – engine to be erected by Messrs Boulton and Watt is completed, and actually works a pump of the same depth of 58 fathoms and 17 inches in diameter at the rate of six strokes of 5½ feet long each in a minute, and consequently making 8640 per 24 hours, it will draw a quantity of water equal to that now drawn by both the present engines, and consequently whatever smaller quantity of coals it uses than 14,080 bushels for 61 days when going at the rate of 6 strokes per minute, will be the real savings in fuel occasioned by the said new engine at that rate of going.
>
> (Signed) James Watt, Matthew Boulton, H. Hawkins Tremayne, Richard Williams, John Williams, Thomas Brown."[9]

By the end of the trial it was established that the Poldice Newcomen engines had achieved a duty of 7,037,800. This figure is slightly lower than Smeaton's average duty of 7.10 millions for the fifteen best Cornish Newcomen engines calculated in 1769 but this may simply reflect the fact that the engines were nearly ten years older at the time of the Poldice trial. To put both Smeaton's and the Poldice figures into context, in 1798 a trial was carried out on twenty-three Cornish Boulton and Watt engines, the average duty being 17,671,000.[10] Each Boulton and Watt engine was monitored and its performance compared with that of the Poldice engines, allowing the Boulton and Watt premiums to be calculated. With regard to the monitoring of performance of the new engines Davies Gilbert writes:

> "The diameter of the various lifting boxes or plungers, the length of the lifts or columns of water, and the lengths of strokes, were matters of common

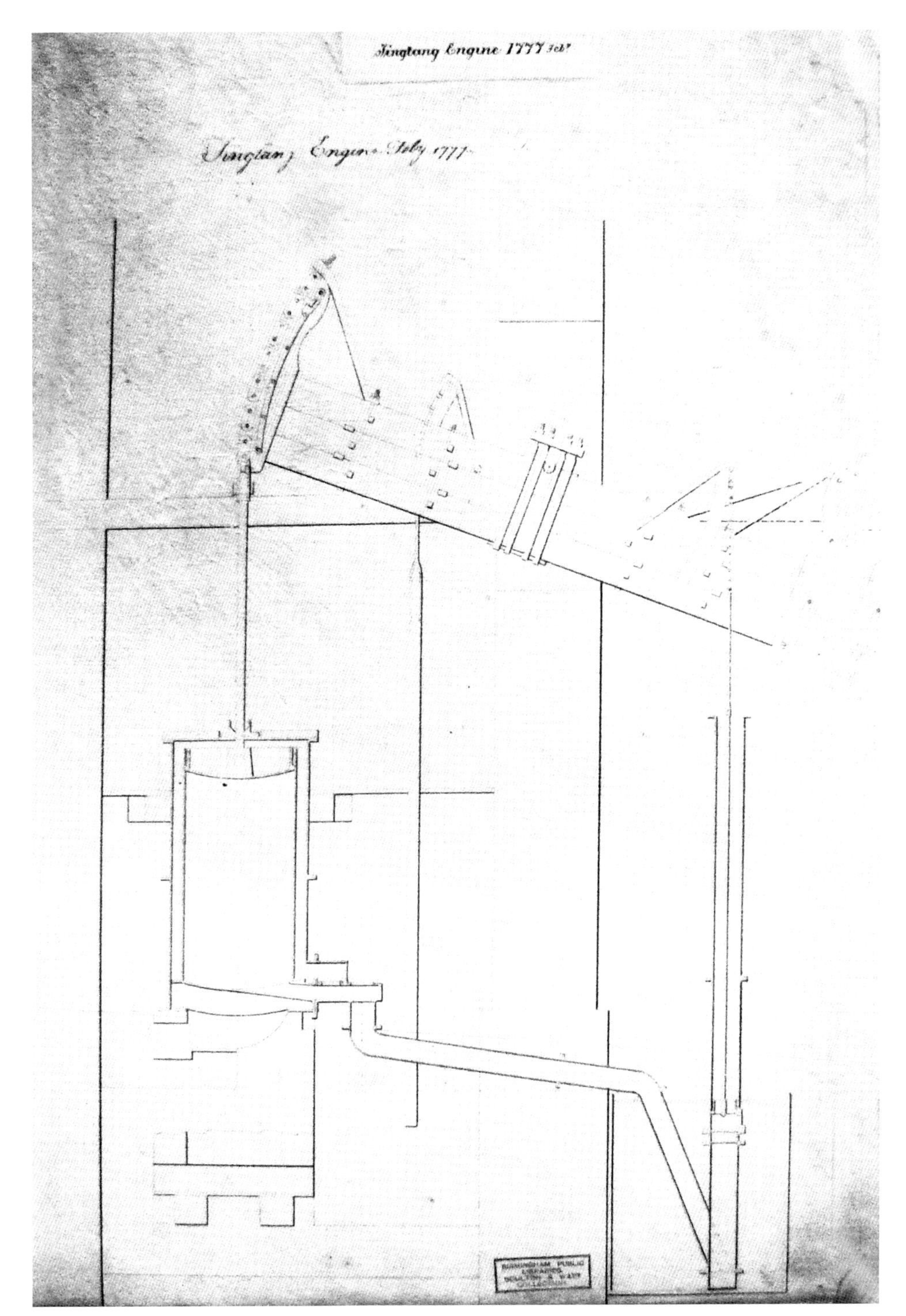

52″ Boulton and Watt engine for Ting Tang mine. This was the first Boulton and Watt engine ordered by a Cornish mine, although, due to delays, not the first started; that honour goes to the 30″ at Chacewater. Reproduced with the permission of the Library of Birmingham, MS 3147/1288.

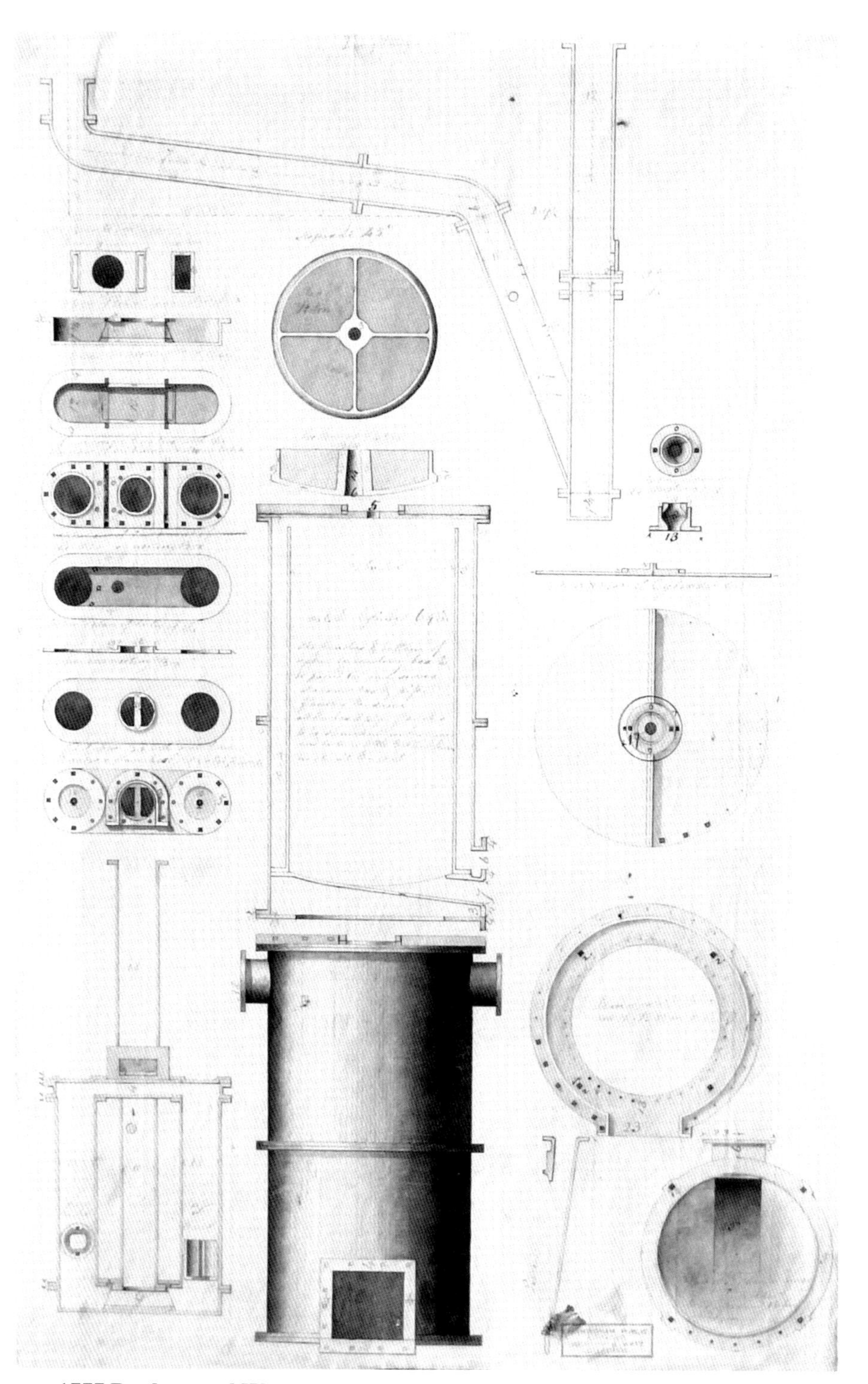

February 1777 Boulton and Watt drawing showing details of parts of the Ting Tang engine. Notable is the jacketing of the cylinder which substantially increased the thermal efficiency of Boulton and Watt's engines in comparison to earlier Newcomen engines. Reproduced with the permission of the Library of Birmingham, MS 3147/1288.

notoriety; and the number of times moved by each engine in a given period was ascertained by a contrivance denominated a Counter, placed on the great beam; this apparatus involves a series of wheels and pinions set in motion by a weight rolling in each direction, and acting through the medium of an escapement similar to that of a clock."[11]

In practice two parallel systems for the charging of engine premiums developed, one system based on performance related savings, the other based on the payment of a fixed price based on the size of the cylinder.[12] How premiums were decided was subject to negotiation between the mine adventurers on the one hand and Boulton and Watt on the other.

The premiums payable to Boulton and Watt would ultimately cause huge discord between them and the Cornish adventurers. However Boulton and Watt were initially welcomed by the Cornish, if not with open arms, at least with a grudging acceptance. Some Cornish engineers, like Jonathan Hornblower (I), seem to have accepted the new engines from the outset. The supposed friction between Jonathan Hornblower (I) and Watt seems to have been largely an invention of nineteenth century writers such as Samuel Smiles; as far as is known Jonathan Hornblower (I) and James Watt remained on cordial terms until the former's death in 1780. Others like John Budge took some persuading; at first Budge appears to have been rather sceptical about Watt's engines. After a visit to the Soho foundry to see one of Watt's engines working, he declined a commission to erect a Watt engine at Wheal Union (Tregurtha Downs). However the two men appear to have become reconciled: In 1779 Watt was very complementary about an engine to his design erected by Budge at Wheal Chance.[13]

Whatever Boulton and Watt's reception, there can be no doubt that their engine gained complete ascendancy over the Newcomen engine within Cornwall remarkably quickly. In 1792 Thomas Wilson, Boulton and Watt's Cornish agent, wrote in regard to the seven Newcomen engines on Wheal Virgin, Wheal Maid and West Wheal Virgin which latterly became part of Consolidated Mines:

"..... in 1779 the seven old engines by which these Mines were then wrought, were unable to keep the Mine drained, so as to work regularly, and that it was necessary to seek the assistance of further Power which (had not Mr Watt before this made his important improvements) must have been of the old construction: this additional power to the Seven Engines then working, would have created such an expense, as, it is evident from experience since, could not have been long borne if attempted; whereas the five engines built by Boulton and Watt, had power sufficient to prosecute the Mine 20 fathoms deeper, and, when that was expended, they by further improvements have enabled the Mine to continue near 10 Years by increasing the power, and reducing the number of Five to Four Engines; viz. by erecting on Wheal Maid one Engine, in the Place of two first built on that Mine, which had got to the extent of their power.

It may not be deemed amiss here to add that Capt. W. Paul, Manager of the Consolidated Mines, made a declaration at a public account in 1783 that he had examined the expense of drawing the water from these Mines, and found the whole (Boulton and Watts savings included) did not amount to so much as it had before cost them, in drawing the water of Wheal Maid only by two of the Seven Engines formerly at work: and this he candidly acknowledged he did in justice to the merits and character of Boulton and Watt."[14]

On May 18th 1783 Watt was able to write to Boulton that they had erected twenty one of their engines in Cornwall and that only one Newcomen engine remained; a major achievement when one considers that it had been slightly less than six years since the first of their engines had been put to work at Chacewater.[15] Boulton and Watt's business supplying new engines to Cornwall would continue strongly until 1788, seventeen new engines being supplied between 1784 and 1787.

Chapter 11 References

1. Pryce W., 1778
2. *Ibid*
3. *Ibid*
4. Dickinson H. W. & Rhys Jenkin, 1927
5. Cited in Hills R. L., 1997
6. Cited in Muirhead J. P., 1854
7. Cited in Dickenson H. W & Rhys Jenkins, 1927
8. Dickinson H. W. & Rhys Jenkin, 1927; Wilson papers: AD1583/10/3 & AD1583/11/69
9. Cited in Gilbert D, 1830
10. Farey J., 1971; Gilbert D., 1830
11. Gilbert D., 1830
12. Dickinson H. W. & Rhys Jenkins, 1927
13. *Ibid*
14. Wilson T., 1792
15. Dickinson H. W. & Rhys Jenkins, 1927

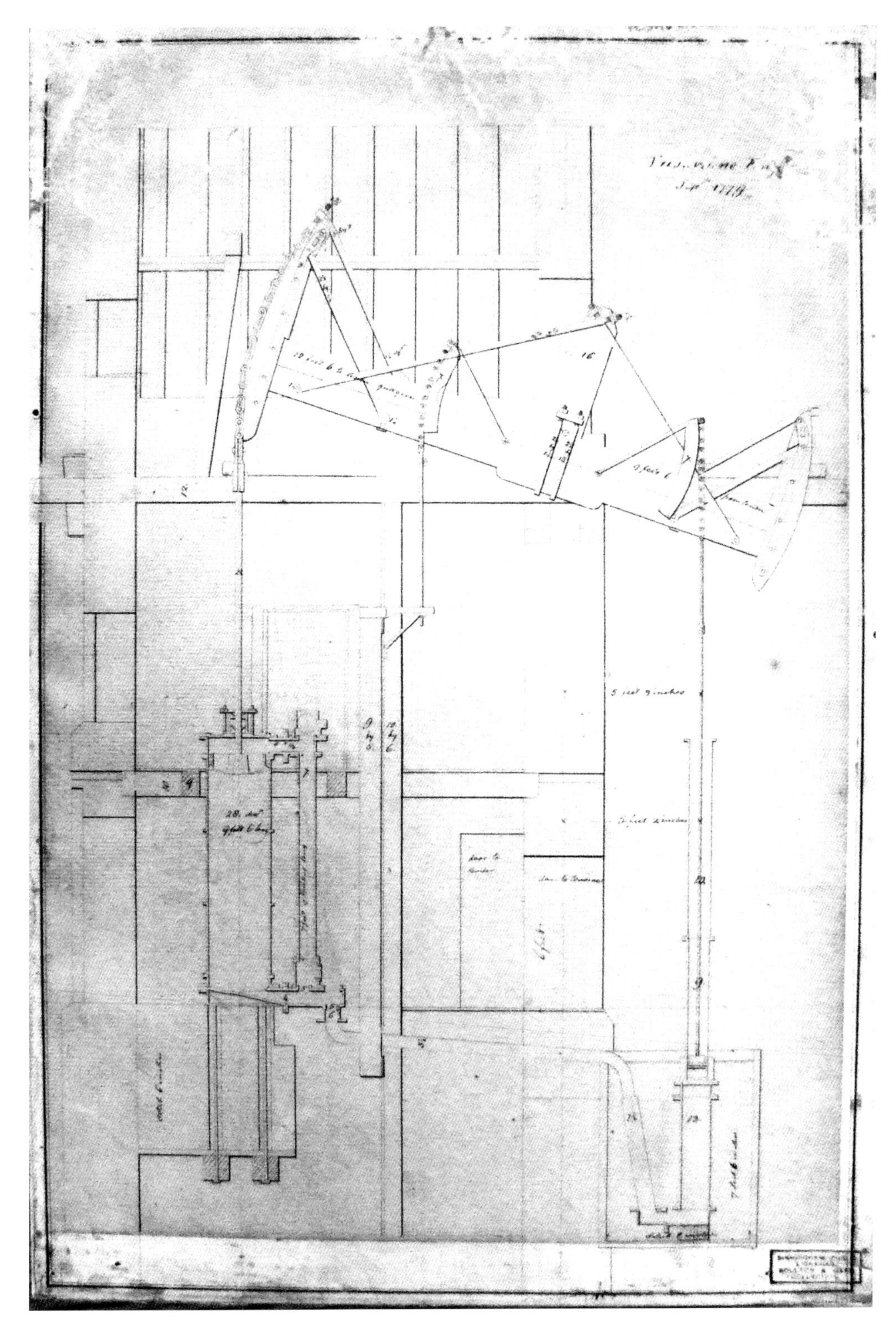

1779 drawing of Boulton and Watt's 28″ engine for Tresavean which was started in August 1780. This engine was moved to Wheal Peevor in 1794. Reproduced with the permission of the Library of Birmingham, MS 3147/1290.

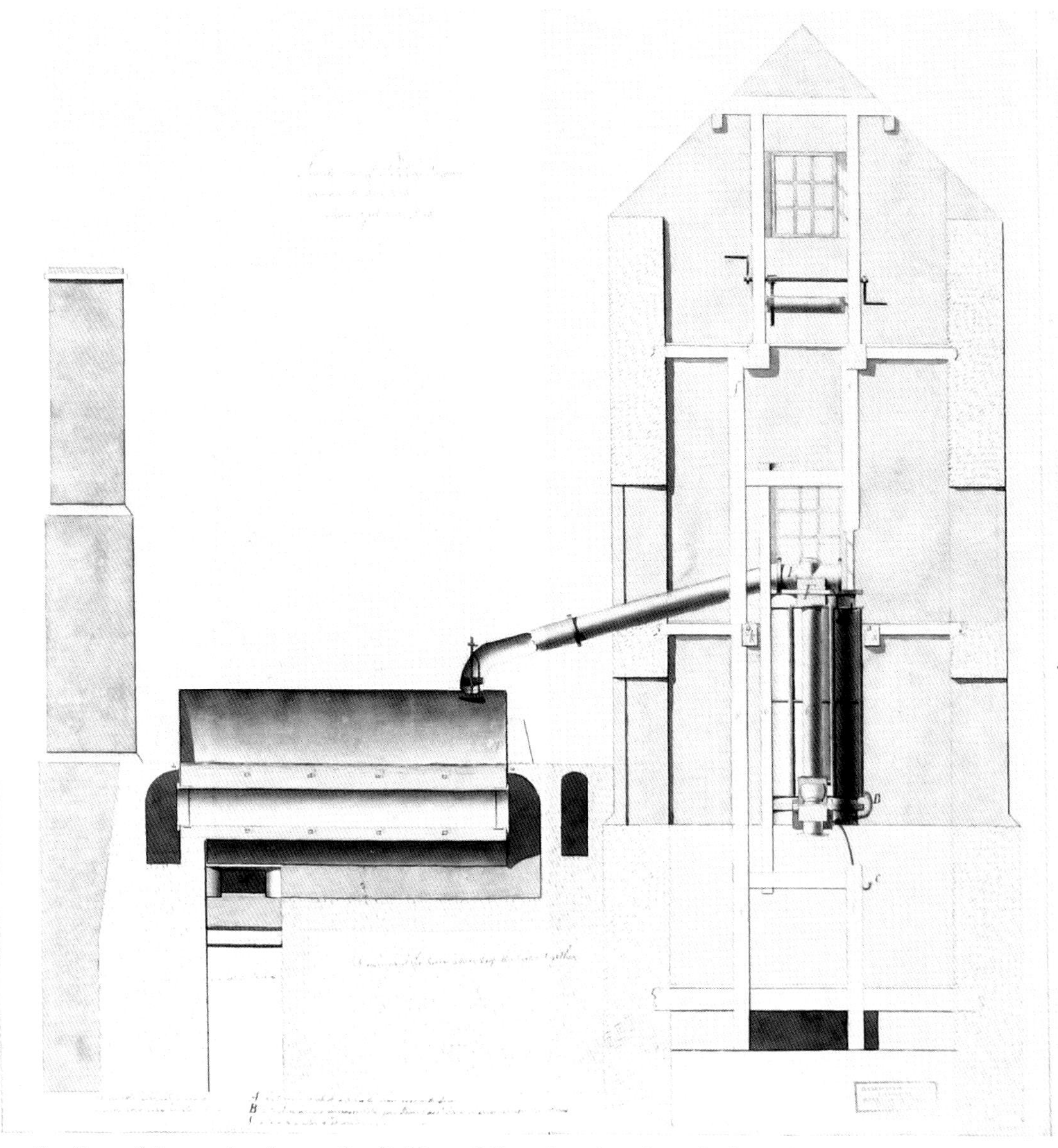

Section of the engine house for Poldory Mine showing the relationship between the boiler and cylinder. Unlike earlier Newcomen type engines, whose boiler was typically located under the cylinder, the boilers of Boulton and Watt engines were offset from the cylinder an, arrangement that would continue on Cornish engines throughout the remainder of the 18th and through the 19th centuries. Reproduced with the permission of the Library of Birmingham, MS 3147/1289.

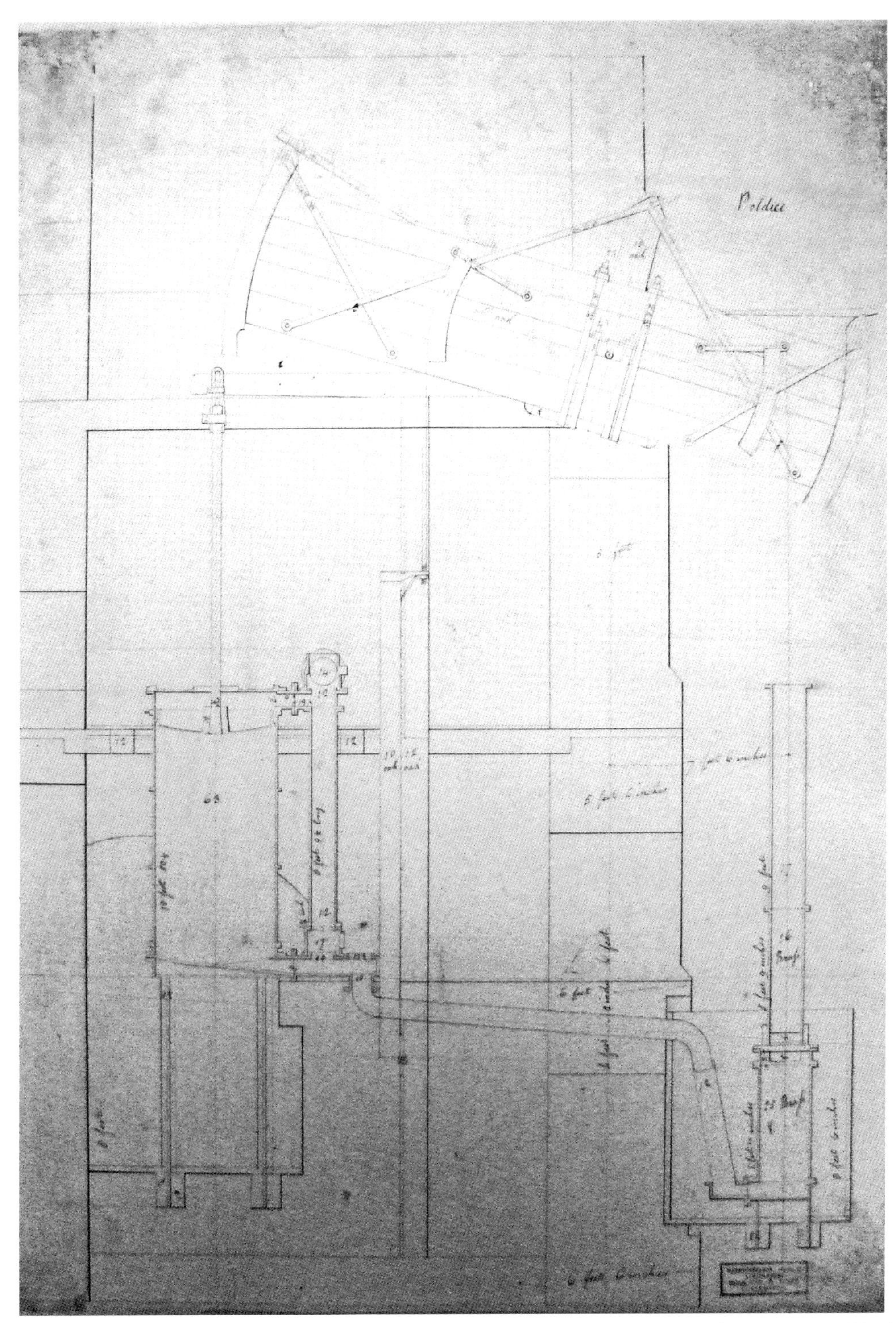

63″ Boulton and Watt engine for Poldice. This is believed to be the mine's Eastern or No.1 engine started in March 1780. This was the first of five Boulton and Watt engines erected on the mine. Reproduced with the permission of the Library of Birmingham, MS 3147/1286.

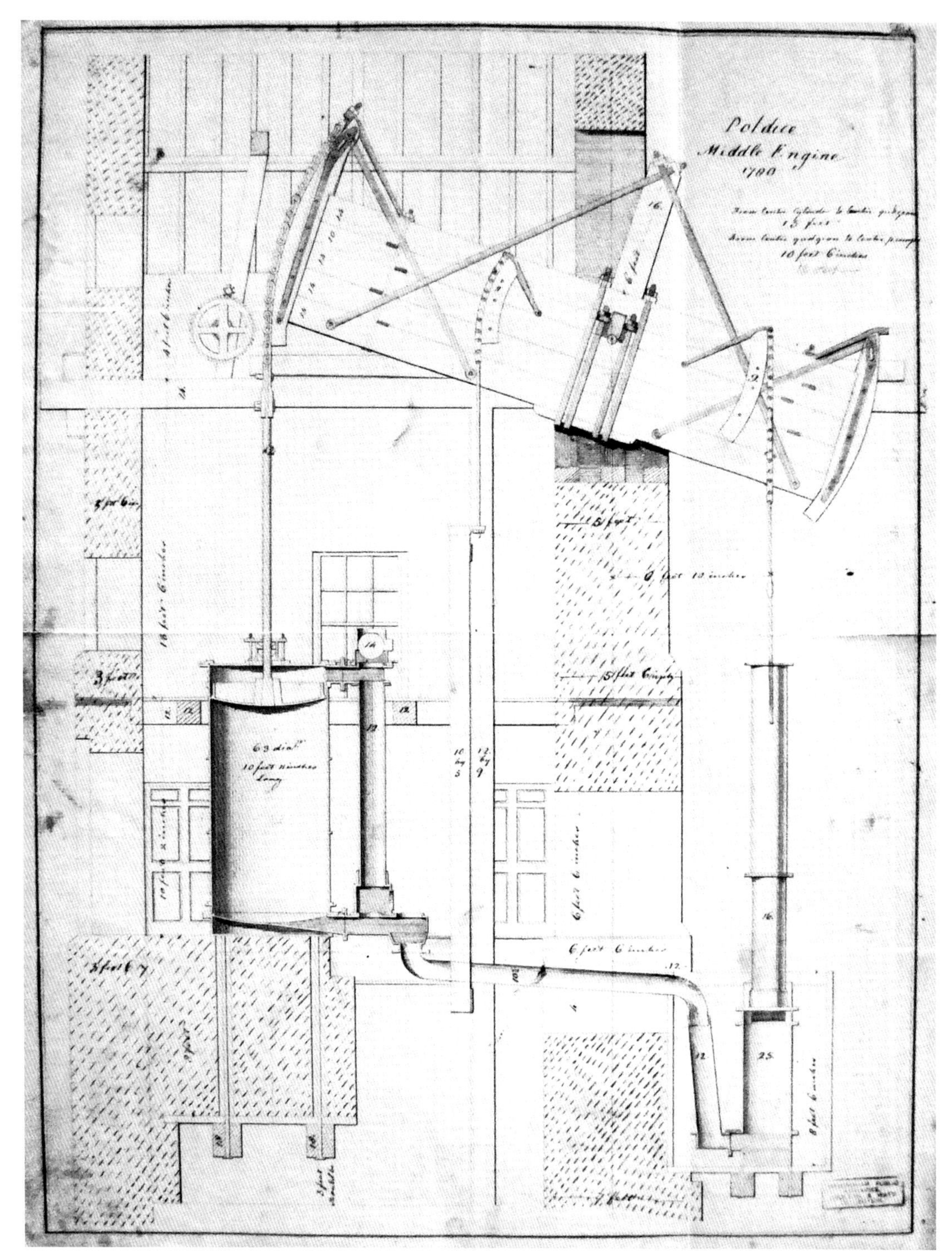

A 1780 drawing of Boulton and Watt 63″ Middle or Western engine for Poldice. This engine was also known as Poldice No.2. It was started in October 1782. The engine was moved to Pednandrea in October 1797.Reproduced with the permission of the Library of Birmingham, MS 3147/1296.

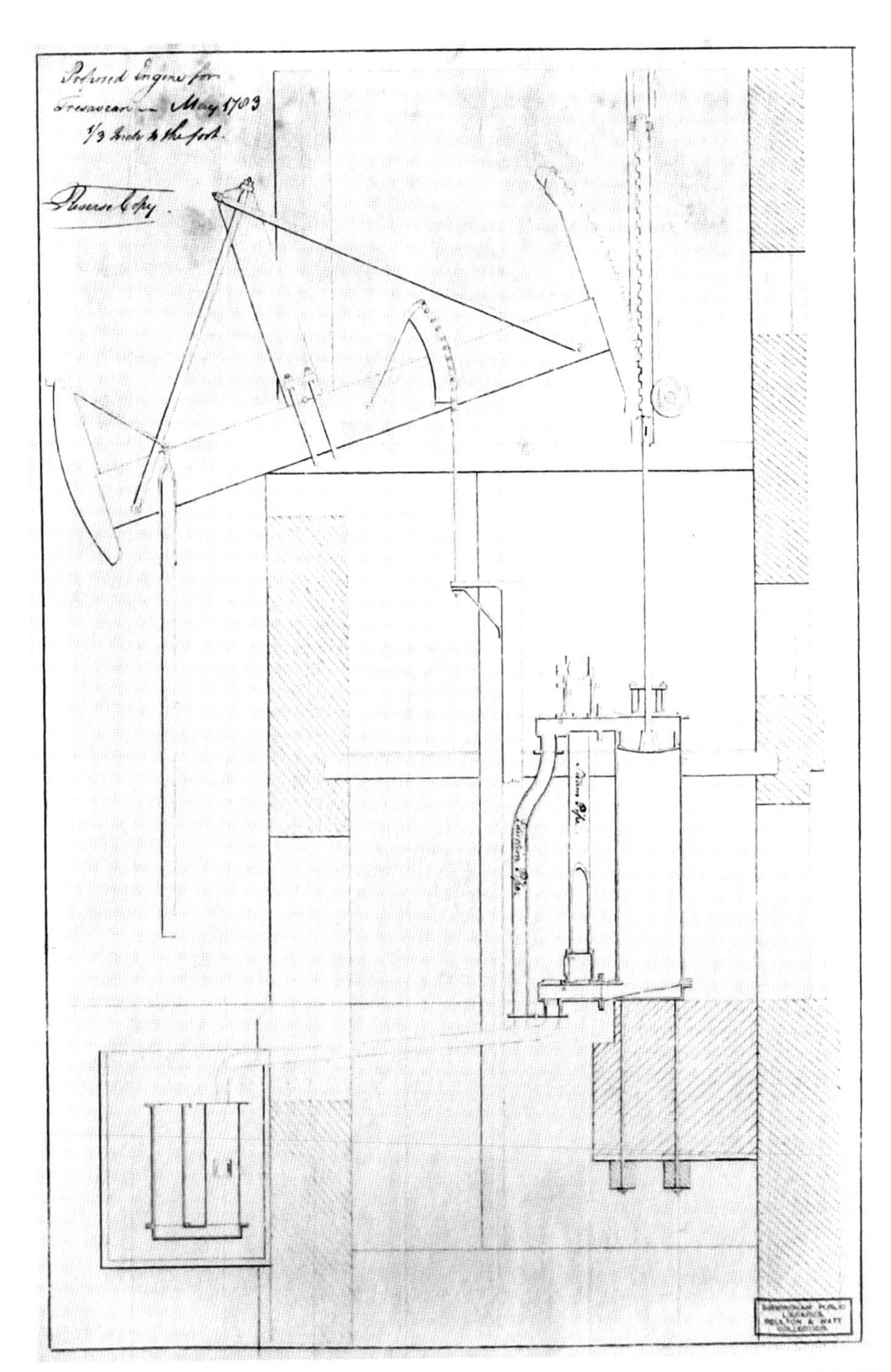

Boulton and Watt drawing of a proposed engine for Tresavean dated May 1783. Of particular interest is the toothed linkage between the arch head and the piston connecting rod. Such crude arrangements would be obviated with the introduction of parallel motion. Reproduced with the permission of the Library of Birmingham, MS 3147/1290.

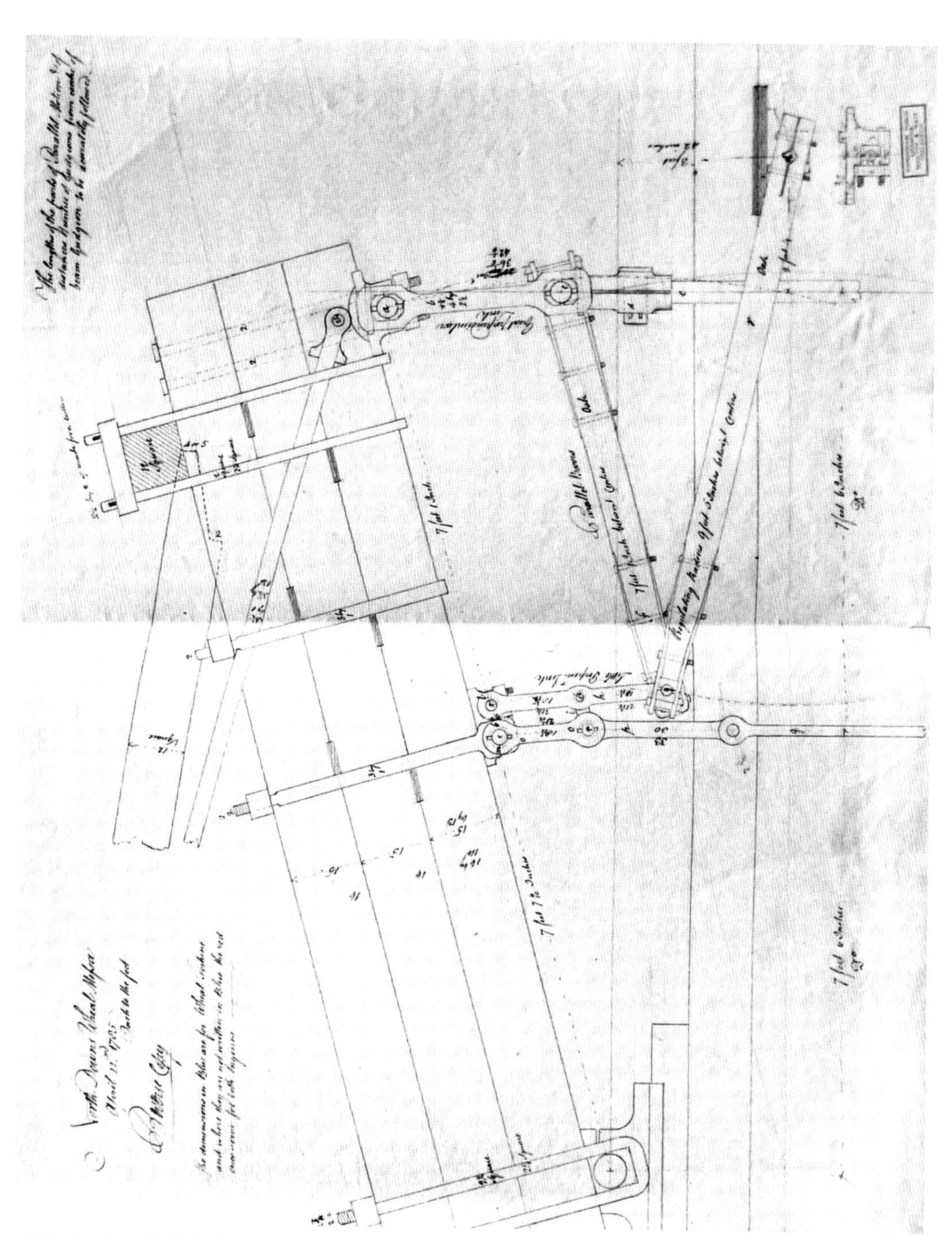

Watt's Parallel Motion. This linkage was patented by James Watt in 1784. It allowed power to be transmitted to a beam on both strokes of the engine which chains attached to an arch head did not. By the nineteenth century parallel motion had rendered chains and arch head beams completely obsolete. Watt's invention made rotary and double acting engines feasible propositions. Reproduced with the permission of the Library of Birmingham, MS 3147/1300

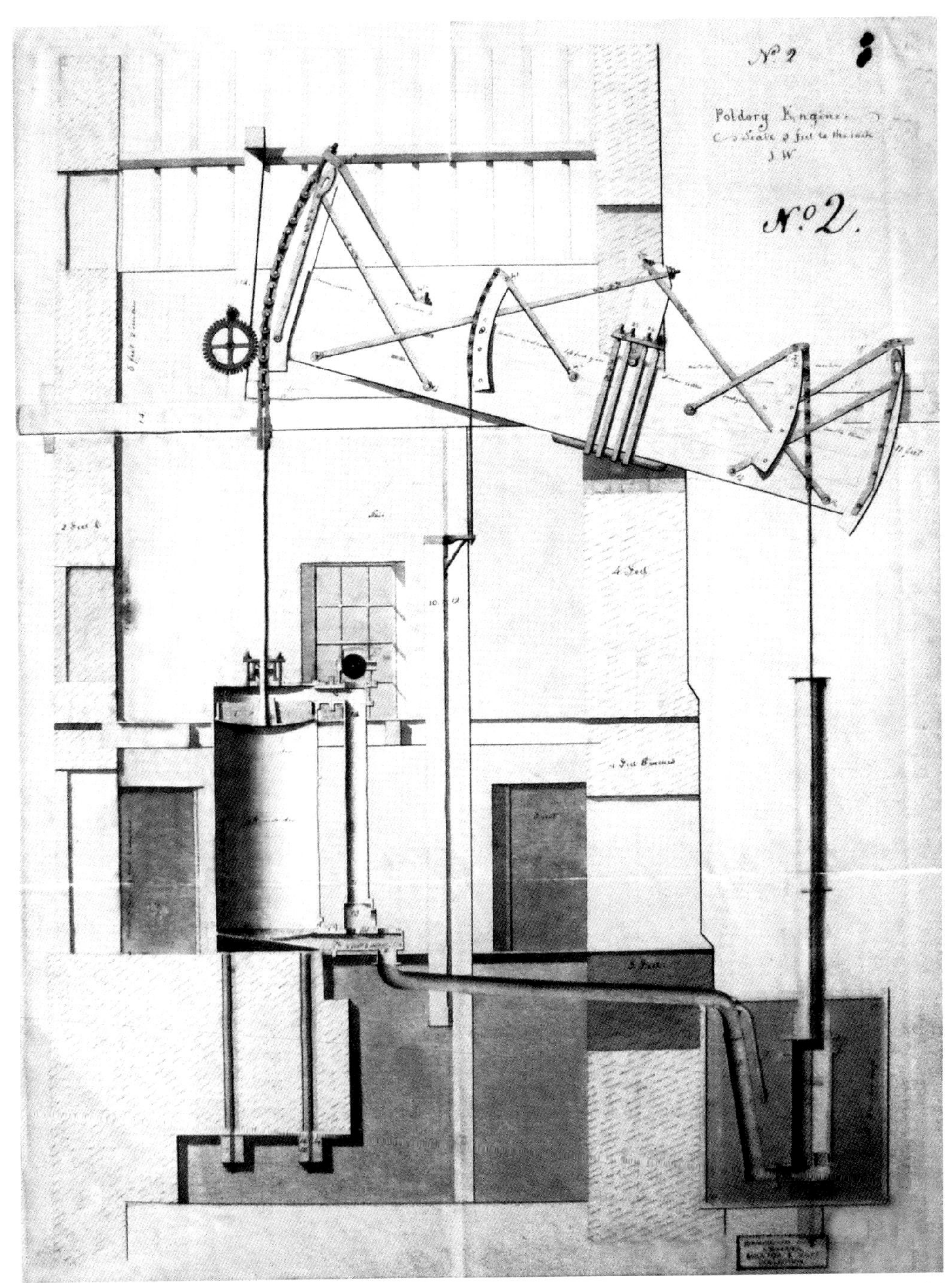

Poldory No. 2 engine by Boulton & Watt. This 45″ engine was started in December 1784 and worked on the Wheal Cupboard section of the mine. Reproduced with the permission of the Library of Birmingham, MS 3147/1325.

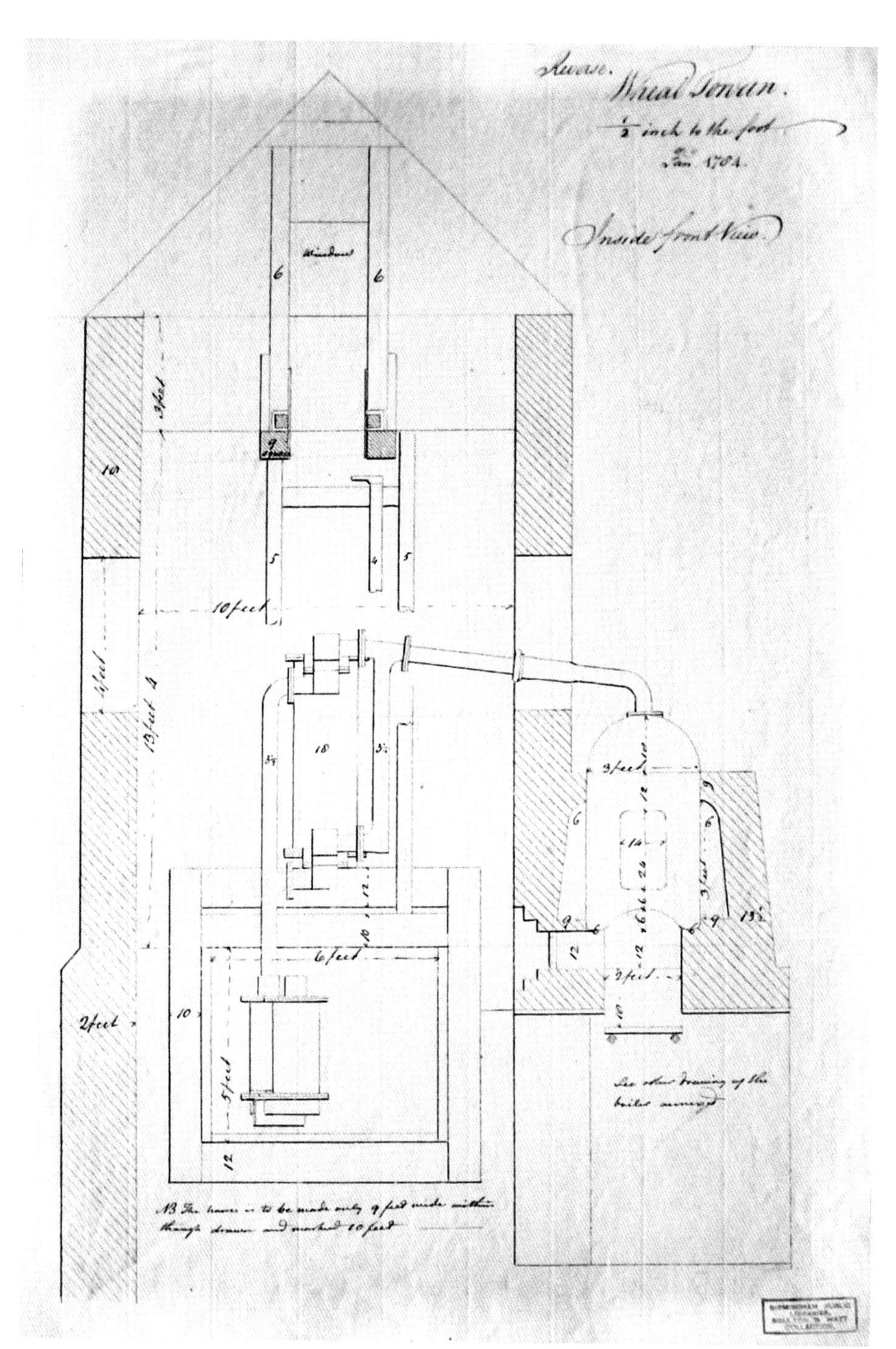

1784 section of Boulton & Watt 18″ engine for Wheal Towan. Although the drawing shows the engine erected in a conventional house, apparently it was erected underground. The engine had a short working life on the mine, working from May 1785 to January 1787, the mine falling victim the ongoing copper crisis. In 1788 the engine had moved to Swansea where it was operating as a rotary engine. Reproduced with the permission of the Library of Birmingham, MS 3147/1302.

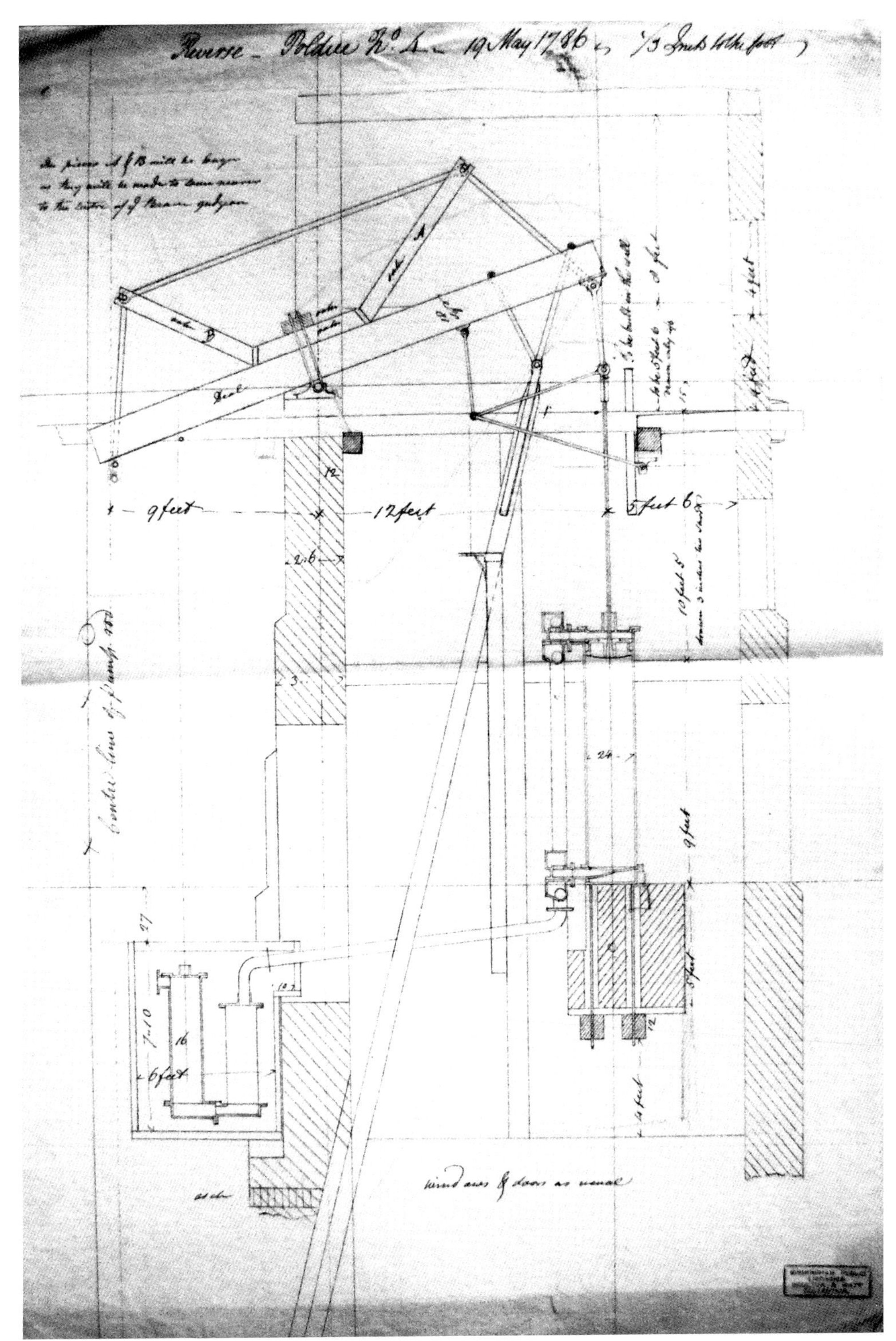

Poldice No. 4 engine. This was a 24″ double engine started in April 1787. The engine last worked at Wheal Gorland. Reproduced with the permission of the Library of Birmingham, MS 3147/1311

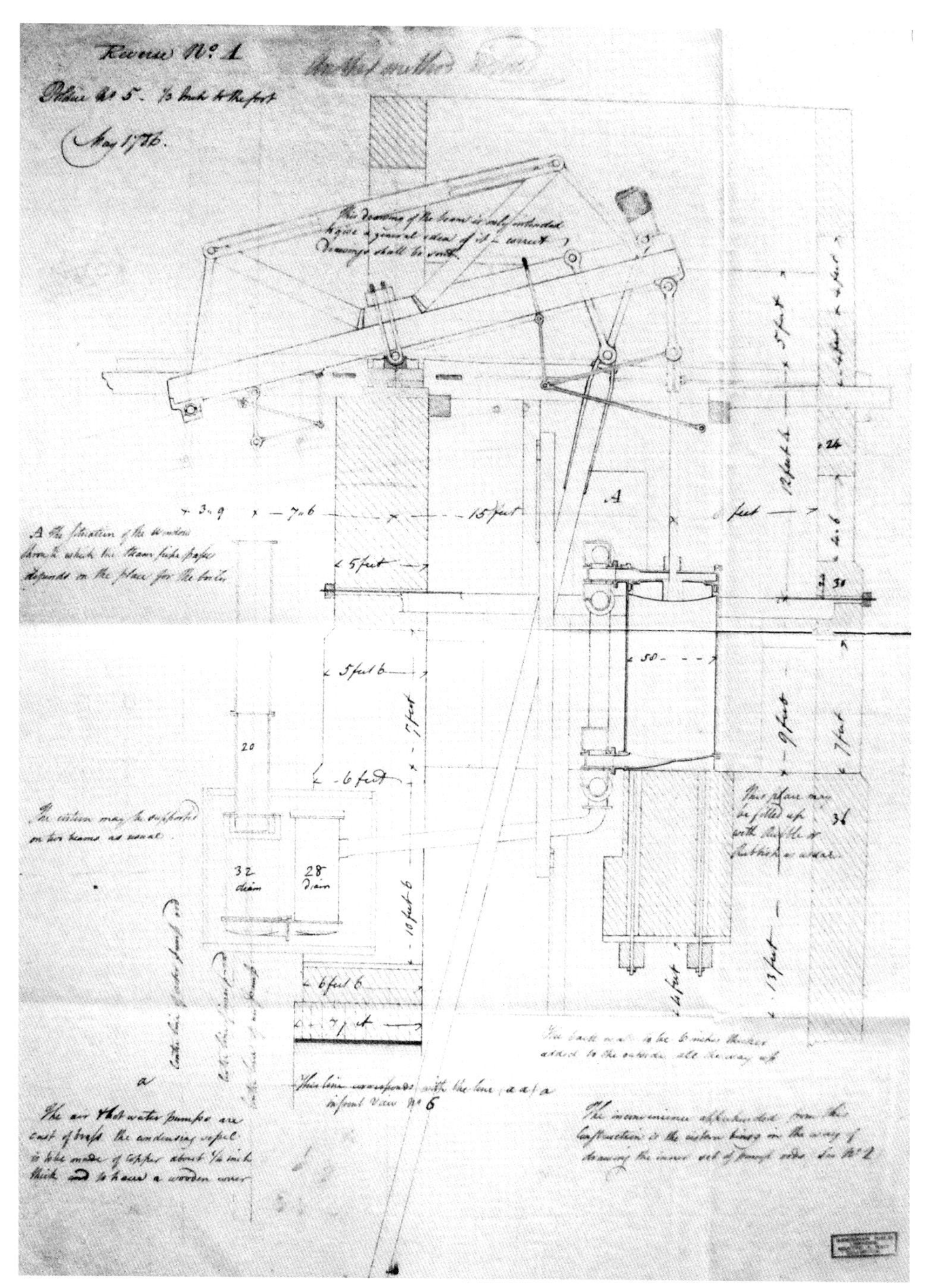

Poldice No. 5 engine. This 58" Boulton and Watt engine started in April 1787. Like the smaller No.4 engine it was a "double engine" hence the use of parallel motion. The engine appears to be designed to drive two sets of pitwork. One set was driven conventionally from the "outside" end of the beam (pump rod not shown on the drawing) whilst a second set was driven from the "inside" end. As lift pumps would have been in use the two sets of pumps would have counterbalanced each other.Reproduced with the permission of the Library of Birmingham, MS 3147/1311.

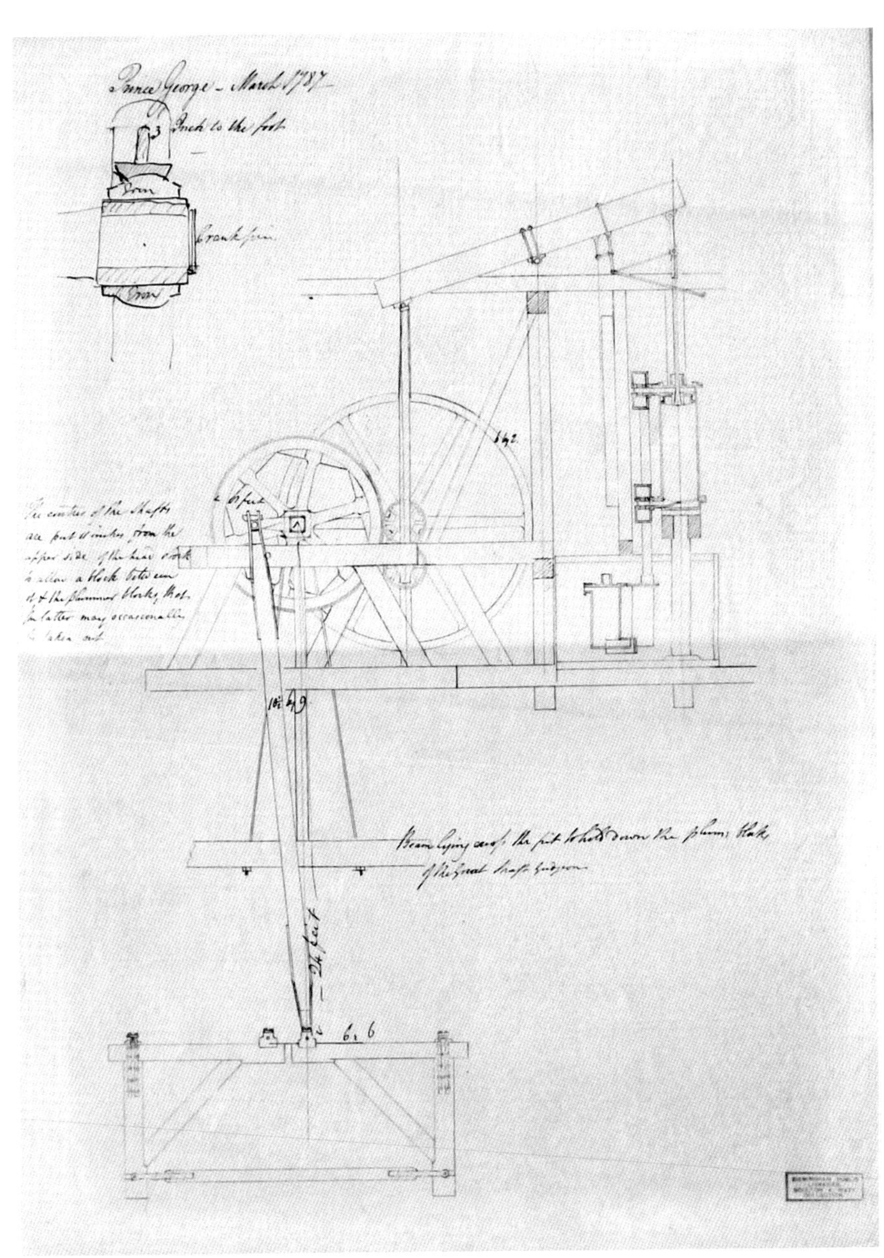

20″ Boulton and Watt engine for Prince George mine. The engine only worked here between February and October 1786. From Prince George the engine moved to Seal Hole mine near St Agnes. The engine was a rotary driving two sets of pit work. Reproduced with the permission of the Library of Birmingham, MS 3147/1282a.

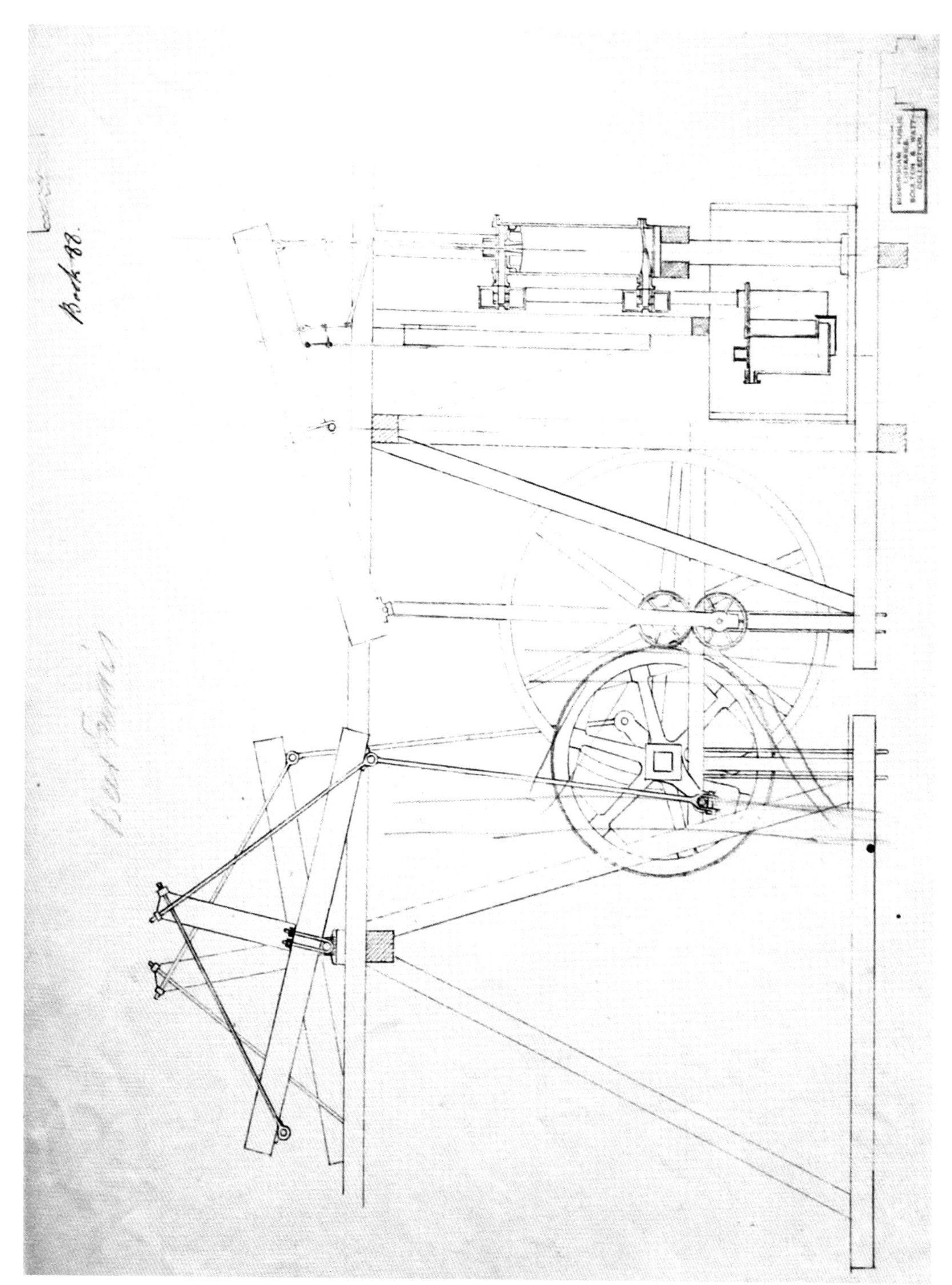

This 20″ Boulton and Watt "double" rotary engine was originally supplied to Wheal Mount, starting work in February 1787, working there until September 1787. In June 1788 the engine was at "Beerferris" Mine on the Devon bank of the River Tamar making it the first mine pumping engine erected in Devon. Reproduced with the permission of the Library of Birmingham, MS 3147/1282b.

Chapter 12
Boulton and Watt in Cornwall:
The 1788 copper crisis to the expiry of Watt's patent in 1800.

The 1780s were a period of crisis for Cornish copper mining; the discovery of a huge easily worked copper deposit at Parys Mountain on Anglesey in 1768 had a profound and ongoing impact on Cornish copper mining. Cornish mines now faced crippling competition. The Anglesey copper mines were controlled by Thomas Williams who also had significant smelting interests. These smelting interests brought Williams into direct conflict with the ring of thirteen Swansea copper smelters, known as the Associated Smelters who dominated the industry. Competition between Williams and the Associated Smelters started to force the copper standard down. The fall in the copper standard was reflected in a fall in the price that the Associated Smelters were willing to pay at the Cornish copper ticketings, potentially spelling disaster for the

Parys Mountain copper mines in 1803. Shallow working depths and proximity to the coast of Anglesey meant that Parys was able to challenge seriously the viability of Cornish copper mining during the 1780s and 1790s. William Havell, 1803.

Cornish mines which had much higher overheads than the shallow Anglesey mines.[1]

In 1785 a cartel, known as the Cornish Metal Company, was set up in opposition to the Associated Smelters. The Cornish Metal Company included Matthew Boulton, a considerable amount of his business being dependent on Cornish mining, John Vivian and John Martyn on behalf of the Cornish mines and Thomas Williams who opposed Associated Smelters and agreed to share the market with the Cornish. The intention was that company would buy ore directly from the mines, smelt it and market it. Strangely the ore itself was smelted by the members of Associated Smelters. By controlling the supply of Cornish and Anglesey ores, the Cornish Metal Company was attempting to establish a monopoly and thus force up the price of copper metal. The monopoly was not however total, with some Cornish and (particularly) Anglesey ore being smelted and sold at lower prices than the Cornish Metal Company. This forced the Cornish Metal Company to drop their price for refined metal. The problem was exacerbated, as the buying arrangement the Cornish Metal Company had agreed with participating mines led to significant overproduction.[2] By June 1787 the situation was becoming critical, on 12th June 1787 James Watt wrote:

> "I understand that W(illia)ms refuses to come to terms of raising fewer ores or making less copper, so that there seems no way left but for the Cornish mines to stop raising so many ores & that can only be done by a fall in the price, which must be a great one, and then everybody left to their discretion those must stop who cannot afford to go on....... On our part we shall surely give our vote for stopping Wheal Towan & other small mines we are concerned in & if our friends the Messrs Foxes could be prevailed on to stop N(orth) D(owns) I think it would be right; though perhaps it may be a proper question for the County at large whether it should be N(orth) D(owns) or Wh(eal) V(irgin) & P(oldice) which should stop for I fear that these latter mines will no longer be profitable under their present expence."[3]

Watt was in particularly low spirits at this time, writing on 26th June 1787 that "I scarcely expect any more Engines will be wanted in Cornwall in our time". He notes that "Anglesea mine is said to be in better state than ever & there is no immediate prospect of W(illia)ms death". Watt concludes: "I foresee times will go hard for us in Cornwall".[4]

In August 1787 North Downs applied to Boulton and Watt for an abatement of their engine dues, a request which was turned down. Watt commented:

> "..... it may be proper to remind them of the great losses they have incurred by mismanagement of the mine at a time when profits might have been gained & which it is unreasonable we should in any shape pay for....... Our only motives for granting abatements has been cases where by continuing the mine there was prospects of repayment, that in this case there was (not) nor is no such

prospect therefore giving up our profits is no more reasonable than it w(ould) be for the merchants to give up theirs or the Captains their salaries."[5]

The situation came to a head at the beginning of April 1788 with the decision to stop production at North Downs.[6] With regard to the situation Matthew Boulton commented that should the mine restart they would forego their dues until such time as the mine was in fork.[7] North Downs was not the only significant mine to close at this time, production also ceasing at Dolcoath. On the 30th April 1788 the 63″ engine at Dolcoath was stopped, the mine not reopening until 1799.[8] The closure of two such important mines hit miners hard, resulting in rioting in April 1788.[9] In return for stopping their mines the North Downs and Dolcoath adventurers received compensation from the surviving mines within the cartel.[10]

From 1788 the demand for new engines dropped off radically, only nine new engines being supplied by Boulton and Watt between 1788 and the end of the century. What did increase was the market for second-hand Boulton and Watt engines, some leading a highly peripatetic existence. Taking the Ting Tang 52″ as an example, the engine was started on the 7th December 1779 and stopped on the 2nd January 1782; it then moved to Scorrier Mine where it was started in September 1782, working until 1st August 1784. From Scorrier the engine then moved to Wheal Fat, starting in May 1785 and working until 2nd November 1786. It then moved to Hallenbeagle starting in October 1787 and stopping on 2nd June 1788. From Hallenbeagle the engine then moved to Wheal Rose.[11] Thus by the late 1780s Boulton and Watt's business in Cornwall had started to change, becoming less dependent on revenue from new engines and more dependent on the collecting of premiums from the mines using their engines, both new and second hand.

Given the real insecurities faced by mines in Cornwall, Boulton and Watt's ongoing demands for engine premiums were becoming extremely unpopular amongst the hard-pressed Cornish adventurers. Surviving correspondence between Boulton and Watt and Thomas Wilson includes numerous requests for abatements on engine premiums. R. A. Daniell, who was a major shareholder in a number of the major Gwennap mines, outlines the problems facing the Consolidated adventurers in a letter dated 26th August 1788:

"At the time the Consolidated mines were sate (*set*) to work, Copper Mining was at a very low Ebb in Cornwall, with a spirit at that time unknown here, The Adv(eventure)rs engaged to lay out nearly £35000 in the undertaking – Your engines were at that time new in the County & the Success which for a while attended the concern was an inducement to other Mines being sate (*set*) to Work, & consequently more fire engines employed, till from an increasing quantity of Copper Ores being brought to the Market the Standard was affected and has consequently reduced this County to its present unhappy situation.

Under the present Circumstances of a very low Standard & a poor Mine it is impossible for the Adventurers in the Consolidated Mines to lay out Considerable Sums in seeking new discoveries, & without such Measures are pursued you are well acquainted with the Event. – The Merchants a considerable time since came forward & readily agreed to a reduction of 5 p(e)r C(en)t on all Goods supplied by them – An Application to you Sir, we thought most likely in the present situation of things to be attended with the desired Effect, & when you consider the Sums of Money lately expended by us, the enormous debt still due from us to Mr. Wilkinson & the Coalbrookdale Co(mpany) taking in the same view the present state of the Mines & when I assure you on my Honor, I never one Moment thought that you meant to claim the Money now uncharged, I flatter myself you will no longer refuse to grant the Favor desired.

I admit the new Engine now built at Wheal Maid to be the most powerful perhaps in the World, but you will please consider the Expences attending the building of it were at least £10000 & had we foreseen the Misfortunes that have attended the Copper Trade it never would have been undertaken – The loss sustained by building so large a Boiler for the Engine wit the consequent Charges you must recollect was very considerable & we understood your premiums for the 2 first Months were given up as a Compensation on the Acc(ount)."[12]

The concerns expressed by the Consolidated adventurers accurately reflected the concerns of mines throughout the county. The question of abatements and levels of premiums remained a bone of contention up to the time of Boulton and Watt's departure from Cornwall at the beginning of the nineteenth century.

Boulton and Watt felt that their patents gave them a virtual monopoly on engine building, not just in Cornwall but throughout the country. Other engineers disagreed, designing and erecting engines to their own designs. In Cornwall Jonathan Hornblower (II) and Edward Bull erected engines during the 1790s and their work is discussed in some detail in following chapters. There were at least two reasons why engines built by "pirates" such as Hornblower and Bull were attractive to Cornish adventurers: Firstly erecting either a Hornblower or a Bull engine, whilst still attracting premium payments, was cheaper than the sums demanded by Boulton and Watt. This made both Hornblower and Bull's engines an attractive proposition to Cornish adventurers. Secondly, and more significantly, it offered the potential to challenge the validity of Watt's patent at law. Both Hornblower, and particularly Bull, became pawns in a wider game with Boulton and Watt on one side and the Cornish adventures (with R. A. Daniell to the fore) on the other.

Boulton and Watt certainly believed that both Hornblower and Bull were "pirates" who were infringing their patents. Of the two, Boulton and Watt were less sure of

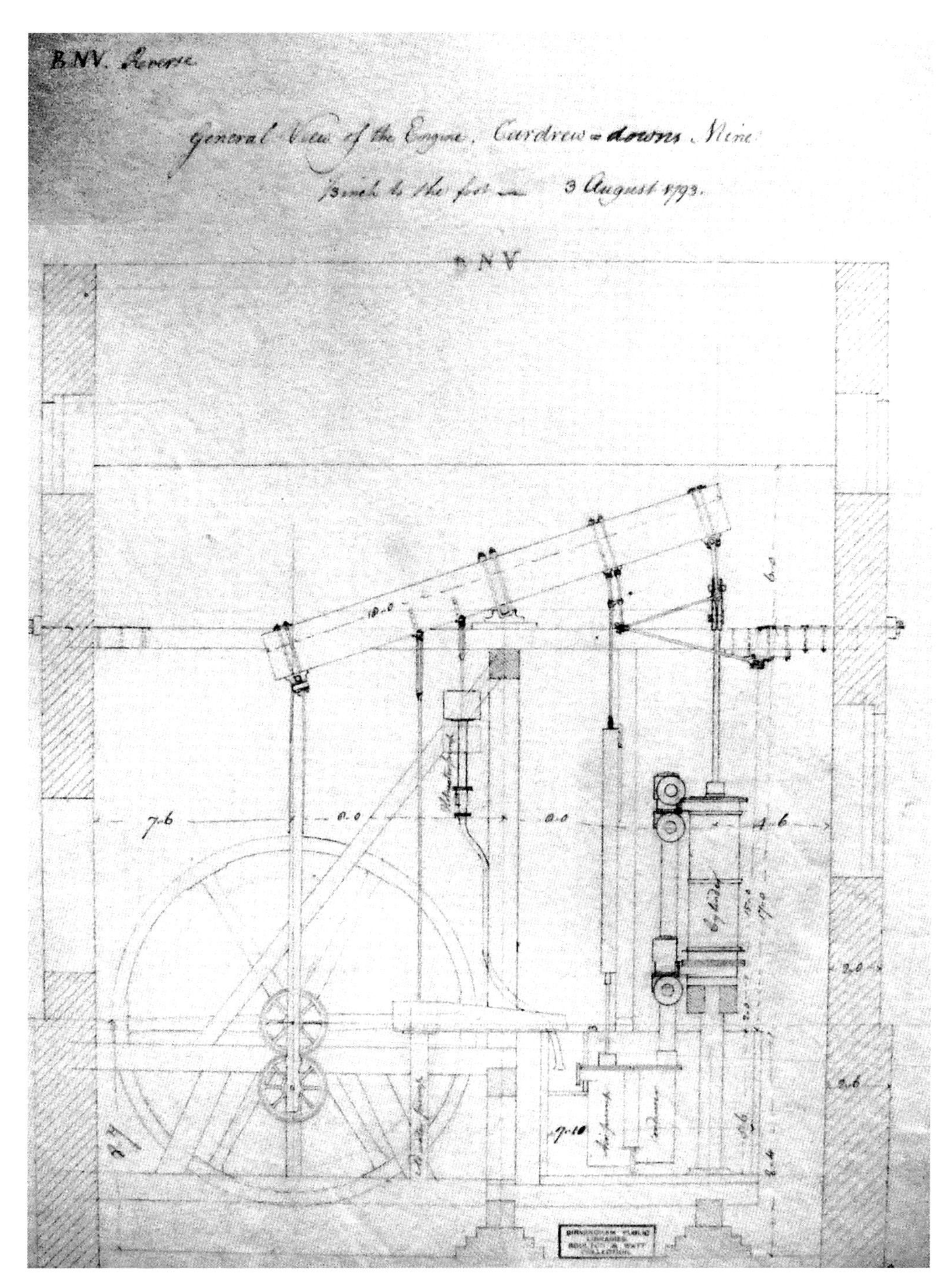

20″ rotary engine which worked at Cardrew Downs between June 1794 and January 1796. The engine came from Nottingham where it had been "burnt by accident". This is typical of their rotary engines of the period. Note the sun and planet gear to convert reciprocating to rotary motion. Reproduced with the permission of the Library of Birmingham, MS 3147/1314.

their ground with Jonathan Hornblower (II). Whilst they regularly fulminated about him in their correspondence, referring to the Hornblower family with disparaging names such as the "Trumpeters" and the "Horners", they never had the confidence to test their claims against Jonathan in a court of law, preferring to proceed by a series of threats.

Edward Bull, though was a different matter. His engine, although having detail differences such as dispensing with the main beam and utilising an inverted cylinder, was in its key essentials a Boulton and Watt engine. Unlike Jonathan Hornblower (II) Boulton and Watt felt confident to initiate legal proceedings against Bull. The decision to take Bull to law court was welcomed by many Cornish adventurers who considered that Watt's patent was too ambiguous to be enforceable and Bull's ongoing legal costs were met by a number of Cornish adventurers. In going to law court Boulton and Watt were looking for two things, first a decision on the part of the court that Bull was infringing Watt's patent and secondly that the patent itself was valid. Bull, and by extension, the Cornish adventurers, wanted to establish that Watt's patent was unenforceable which would allow Bull to continue erecting engines without hindrance and would free the adventurers from the yoke of Boulton and Watt's engine premiums.

The case of Boulton and Watt versus Bull was tried in the Court of Common Pleas on 22nd June 1793. Neither side got what it wanted. Whilst the court agreed that Bull had infringed Watt's patent, the decision as to whether the patent was valid remained undecided. This partial decision did give Boulton and Watt the confidence to take out an injunction on 22nd March 1794 preventing Bull from erecting any more engines which infringed Watt's patent. The case returned to the Court of Common Pleas in 1794 – 1795. As with the 1793 case the outcome was ambiguous: On 16th May 1795 the court offered its judgement, two of the judges finding in favour of the patent and two against. In spite of the court's failure to reach a definitive answer, the March 1794 injunction was upheld by Lord Rosslyn to the serious detriment of Bull's practice.[13]

It was essential for Boulton and Watt to establish the validity of Watt's patent if they wished to continue collecting premium payments from Cornish adventurers, many of whom were refusing to pay the requested premiums. As an example in January 1794 United Mines "determined to withhold all payments claimed by Boulton and Watt"....."until Boulton and Watt's patent was established".[15] By the end of August 1796 outstanding engine premiums amounted to £22,221.[16] Mines in debt to Boulton and Watt at this time included:

Consolidated Mines	£4095 0s 0d
United Mines	£6100 0s 0d
Wheal Rose	£127 13s 4d

Wheal Crenver	£4885 5s 4d
Wheal Leeds	£166 5s 0d
Hallamaning	£649 2s 8d [17]

The next phase in Boulton and Watt's campaign did not directly involve Cornwall although it had a Cornish connection in that it involved Jabez Carter Hornblower (1744 – 1814), elder brother of Jonathan Hornblower (II). Jabez had at one time worked as an engine erector for Boulton and Watt, before working with his brother Jonathan. By 1790 Jabez had established himself in London as a manufacturer of steam engines and other machinery. He was working in partnership with J. A. Maberley who had purchased Isaac Mainwaring's engine patent of 1791. Jabez improved on Mainwaring's engine and began erecting them. As built these engine used cylinder covers and a separate condenser, both of which were covered by Watt's patent of 1769. In January 1796 Boulton and Watt took out an injunction against Hornblower and Maberley, preventing them from erecting further engines to this design. Hornblower and Maberley approached Boulton and Watt with a view to gaining a licence to continue erecting their engines, without success. By the spring of the same year Hornblower and Maberley were trying to get the injunction overturned in Chancery. The case was heard on Thursday 2nd June 1796 and much to Hornblower and Maberley's "surprize and vexation" the Chancellor upheld the injunction "in its fullest extent".[18] Defying Boulton and Watt and in breach of the injunction Hornblower and Maberley contracted to erect an engine at the Meux Brewery in London in 1796. By June 1796 Boulton and Watt decided to take further legal action against Hornblower and Maberley. Whilst on the surface this appeared a simple case of engine piracy James Watt felt that there were sinister Cornish undertones:

> "We find ourselves at last constrained to bring an action at Law ag(ainst) Hornblower and Maberley who we have reason to believe are supported by Danie(ll) & all his infernal crew."[19]

Having failed to get their decision in *Boulton and Watt v Bull*, the Cornish adventurers were now supporting Hornblower and Maberley with a similar aim in mind.[20]

The case was heard in the Court of Common Pleas in December 1796 and it found in favour of Boulton and Watt. As a result Maberley "discharged his men & broke up his establishment" and it was noted that "this engineering business will have cost him first and last, not less than £8000". James Watt Junior reported that Maberley intended to pursue the matter as far as the House of Lords if necessary.[21] The outcome of the case left James Watt Junior in gloating mood:

"Maberley, it seems, grows daily less presumptuous, but we presume maintains his Intention of carrying us on to the House of Lords – We shall lead him such a dance as he has but little idea of, although he might for some faint idea of the future from the past, if he would allow himself time to reflect upon his misfortunes. – I rejoice exceedingly that Daniell is in for part of Maberley's expenses. I wish only that he had to pay the whole."[22]

Maberley did appeal against the decision of the Court of Common Pleas. The case came before the King's Bench on the 16th November 1798, when arguments were heard and it was ordered that the hearing be deferred to the following (Hilary) term.[23] The hearing was reconvened on 25th January 1799 and it found in favour of Boulton and Watt: Watt's 1769 patent was valid in law.[24] The Boulton and Watt camp was ecstatic, James Watt Junior writing on the day of the court's decision:

"Send forth your Trumpeters & let it be proclaimed in Judah that the Great Ninevah has fallen; let the Land be clothed in Sackcloth & in Ashes! Tell it in Gath, and speak it in the streets of Ascalon. Maberley and all his host are put to flight!.......... We shall now see, what new devise, these villains who have hitherto cloaked their dishonour under the cover of quibble, of Dry Law, can resort to, to save them from the pecuniary mulet (Author's note: *presumably mallet*) which awaits their misdeeds. They shall now render unto Caesar, the things which are Caeser's, and the protraction of their punishment shall only render it the more compleat....."[25]

With the decision on their side Boulton and Watt lost no time in attempting to recover outstanding premiums. On 29th January 1799 James Watt Junior wrote to Thomas Wilson requesting details of the "Recusant Mines".[26] Wilson acted quickly sending the requested details on 2nd February. Wilson's list included the following mines:

Wheal Treasure	£1596 10s 0d
Wheal Gons	£3685 0s 0d
Godolphin	£945 0s 0d
Wheal Crenver	£8323 13s 0d
United Mines	£9833 6s 8d
Consolidated Mines	£13958 6s 8d
Hallamaning	£649 2s 0d
Wheal Abraham	£392 14s 0d
Balcoath	£94 4s 11d
Wheal Rose	£187 13s 4d

Carzize Wood	£318 0s 4d
Wheal Anne	£592 0s ½d

According to Wilson's calculations recusant mines owed Boulton and Watt £41933 5s in outstanding premiums.[27] In his covering letter Wilson records details of a chance meeting with R..A. Daniell, arguably Boulton and Watt's most implacable foe in Cornwall:

"Since my last I met Mr Daniell in the Street, who address'd me very mildly on the subject of your Victory said he had been much deceiv'd, & seemed anxious to learn what orders I had rec(eive)d I told him none; and we parted with my promising to inform him as soon as I had; he had previously told Edwards the decision had not given him a moments Pain, but if I may judge from his appearance, he had been, & was then very uneasy."[28]

Wilson's assessment of Daniell's state of mind was probably fairly accurate given his extensive holdings in the recusant mines. In May 1799 Daniell was the major shareholder in both United Mines in which he held 1/6th share and Consolidated Mines in which he held 29/64th share.[29] Pole records that:

"in the year of the settlement, Messrs Boulton and Watt received about £40,000 for arrears of patent dues. Messrs Daniel, the agents for adventurers in some works forming part of the Consolidated Mines, paid in one draft the sum of £16,000."[30]

With the expiry of Watt's patent in 1800, Boulton and Watt systematically disengaged from Cornwall, pursuing outstanding premium arrears and disposing of their shares in Cornish mines. The legacy they left behind them is a conflicting one. Certainly without the Boulton and Watt engine, deep mining in the county would have been drowned out by the end of the 1770s. Conversely the issue of engine premiums may have threatened the financial viability of mines, particularly during the hard times of the late 1780s and early 1790s. This was certainly the opinion of many Cornish adventurers; however how much of this was objective assessment as opposed to subjective expression of self-interest is a moot point. It has also been argued that the all embracing nature of Watt's 1769 patent stifled engine development until its expiry in 1800. In support of this argument one could cite Boulton and Watt's ongoing battle with Jonathan Hornblower (II). However, were this the case, one would expect a golden age of Cornish engine building directly after the patent expired. In fact the opposite happened, with the departure of Boulton and Watt and skilled men like Murdock, Cornish engineering descended into what Matthew Loam, writing in 1859, described as "the dark age..... of Engine improvement", not recovering until Arthur Woolf (1766 – 1837) returned to the county in 1811.[31] Of this "dark age" Pole comments:

> "In 1800, Messrs. Boulton and Watt's patent having expired, their connection with the mining district ceased, and was never afterwards renewed.
>
> The agents and assistants who had acted under their direction, and had been in the habit of managing their engines, were then recalled from the county, and their places were supplied by others who had neither the experience nor the scientific acquirements requisite to keep the engines in the same good order and to work them to the same advantage as formerly.
>
> At the same time, other makers, who had long before been excluded, began to manufacture and erect the engines; but being ignorant of the principles which had guided the patentees in their construction, and little accustomed accuracy of workmanship, their attempts, with the exception of some few made by Jonathan Hornblower, produced very indifferent machines, much inferior in regard to economy to any of those which they were imitations."[32]

Whilst, particularly from a Cornish perspective, it would be easy to develop a convincing critique of Boulton and Watt's activities in the county the counter arguments should perhaps hold greater sway. No one can seriously dispute that Watt's design was good and that the separate condenser was a stroke of genius. Boulton and Watt's engines proved to be the saviour of Cornish deep mining in the late eighteenth century. The much vaunted "Cornish" beam engine of the nineteenth century was firmly based on the principles of the Watt engine but worked at higher pressure, using steam expansively.

Chapter 12 References

1. Pennington R. R., 1977
2. *Ibid*
3. Wilson papers: AD1583/2/54
4. Wilson papers: AD1583/2/58
5. Wilson papers AD 1583/2/71
6. Wilson Papers AD1583/3/17
7. Wilson papers: AD1583/2/73
8. Buckley J. 2010; Wilson papers: AD1583/10/3 & AD1583/11/69
9. Wilson papers: AD1583/3/18
10. Barton D. B., 1961, Buckley J., 2010, Pennington R. R., 1977
11. Wilson papers: AD 1583/10/3 & AD1583/11/69
12. Wilson Papers: AD1583/3/39/2
13. Davies J., 1816; Dickinson H. W. & Jenkins R., 1927; Farey J., 1827; Muirhead J. H., 1859; Wilson papers: AD1583/7/19
14. Dickinson H. W. & Jenkin R., 1927
15. Wilson papers: AD1583/11/101

16. Pennington R. R., 1975
17. Wilson papers: AD1583/11/92
18. Wilson papers: AD 1583/9/30
19. Wilson papers: AD 1583/9/33
20. Dickinson H. W. & Jenkins R., 1927; Farey J., 1827; Harris T. R., 1976; Tann J, 1981
21. Wilson papers AD/1583/9/55
22. Wilson papers: AD1583/9/56
23. Wilson papers AD 1583/10/85
24. Dunford C. & Hyde East E., 1827, Wilson papers: AD1583/11/2
25. Wilson papers: AD1583/11/2
26. Wilson papers: AD1583/11/3
27. Wilson papers: AD 1583/11/98
28. *Ibid*
29. The Lord Hawkesbury, 1799
30. Pole W., 1844
31. Harris T. R., 1966, MacKay D., 2010
32. Pole W., 1844

Chapter 13
List of Boulton and Watt engines erected in Cornwall.

This list is largely based on a list dated 3rd May 1798 in the Wilson papers (Wilson papers: AD1583/10/3). AD1583/11/69 has also been extensively used. Similar lists, albeit differing in detail, may be found in Dickinson and Jenkin's *James Watt and the steam engine* of 1927 and Jennifer Tann's article in the 1995 – 1996 *Transactions of the Newcomen Society*.

Mine	Date	Cylinder	Notes	References
1. Chacewater	September 1777	30″	1. Stopped 1st June 1778 2. To Scorrier Mine, started October 1778, stopped September 1782 3. To Wheal Virgin St. Hilary, started December 1782, stopped July 1785 4. To Crane Mine, started April 1786, stopped 30th June 1787 5. To Poldory, 1791 stopped 1794 6. To Wheal Susan.	Wilson papers
2. Chacewater	Started 1st August 1778	63″	Used parts of Smeaton's 72″; worn out 1783.	Wilson papers
3. Hallamaning	Started 21st July 1779	40″	1. Stopped 15th January 1786 2. To Retallack, started August 1787, stopped December 1787 3. To Wheal Treasury, started 1st July 1790, stopped December 1793	Wilson papers
4. Ting Tang	Started 7th December 1779	52″	1. Stopped 2nd January 1782 2. To Scorrier Mine, started September 1782, stopped 1st August 1784 3. To Wheal Fat, started May 1785, stopped 2nd November 1786 4. To Hallenbeagle, started October 1787, stopped 2nd June 1788 5) To Wheal Rose	Wilson papers

5. Wheal Union	Started 13th December 1779	63″	1. Stopped 31st August 1780 2. To Kestal Adit, started 1st July 1781, stopped 2nd June 1784 3. To North Downs (Wheal Hawk), started May 1786, stopped 31st May 1788 4. To North Downs (Wheal Rose) started November 1792	Wilson papers
6. Poldice, No. 1 Engine a.k.a Eastern Engine	Started March 1780	63″	1. Stopped 1795 2. To Wheal Treasure (working on "Bull's plan")	Wilson papers
7. Wheal Chance	Started April 1780	63″	1. Stopped 2nd June 1783 2. To Polgooth, started 1st March 1784, stopped 1796	Wilson papers
8. Ale and Cakes	Started 1st July 1780	58″		Wilson papers
9. Poldory	Started 1st July 1780	48″	Stopped December 1794	Wilson papers
10. Tresavean	Started August 1780	28″	1. Stopped 31st January 1787 2. To Wheal Peevor 1794	Wilson papers
11. Wheal Treasure / Treasurey	Started 6th November 1780	36″	1. Stopped 1st March 1783. 2. To Wheal Gons, started 9th November 1784, stopped October 1788 3. To Wheal Crenver, started June 1789, assumed stopped May 1793	Wilson papers
12. Dolcoath	Started 9th August 1781	63″	1. Stopped 30th April 1788 2. To Wheal Gons, started 1st May 1789	Wilson papers
13. Wheal Crenver	Started 19th February 1782	48″	1. Stopped January 1786 2. To Godolphin, started 1st July 1787, stopped December 1794 3) To Herland, started June 1795	Wilson papers
14. Pool	Started 1st April 1782	60″	1. Stopped 2nd April 1784 2. To Chacewater, started May 1785, stopped 31st March 1789 3. To Poldory 1794	Wilson papers
15. Trevaskus	Started 9th August 1782	45″	1. Stopped November 1784 2. To North Downs (Wheal Rose), started 1st March 1786 3. To Hallenbeagle, started November 1792. 4. To New Briggan, started December 1797	Wilson papers

16. Poldice, No. 2 Engine, a.k.a. Western Engine	Started October 1782	63″	1a. Circa 1789 received new cylinder ex. Poldice No. 3 engine (see 9 below) 1b. Stopped? 2. To Pednandrea, started October 1797	Wilson papers
17. Consolidated Mines Wheal Virgin Elvan	Started October 1782	58″	Used by Bull as part of double cylinder engine of 1789; returned to former state 1790; stopped September 1797	Wilson papers
18. Consolidated Mines Wheal Virgin East	Started October 1782	56″	Stopped To West Wheal Virgin 1798	Wilson papers
19. Wheal Maid East	Started October 1782	58″	To Wheal Virgin East in place of 21 (above), started September 1797	Wilson papers
20. Wheal Maid West	Started October 1782	50″	Moved to same house as Wheal Virgin Elvan engine (see 17 above); used by Bull as part of Double cylinder engine of 1789, returned to former state 1790; idle May 1798	Wilson papers
21. West Wheal Virgin	Started October 1782	52″		Wilson papers
22. Wheal Maid Whim	Started May 1784	18″	Rotative 1. Stopped September 1793 2. To Foxes & Neath Abbey for a coal work – probably not used 3. To Herland, started May 1797	Wilson papers
23. Crane	Started 20th November 1784	14¾″	Double engine 1. Stopped September 1785 2. To Chacewater, converted to rotary/ whim engine; started February (?) 1787; stopped March 1789 3. To Wheal Carpenter; started August 1793 stopped December 1796 4. To United Mines; started April 1797	Wilson papers
24. Poldory, New Engine	Started December 1784	45″	"Working at Cupboard from its first erection"	Wilson papers
25. Wheal Towan	Started May 1785	18″	Double engine 1. Working underground at Wheal Towan. Stopped January 1787 2. To Mr Morris, Swansea, 1788; made rotative for drawing coal	Wilson papers

26. Poldice, No. 3 Engine	Started June 1785	63″	1. Stopped 30th April 1787 2. To Chacewater, started July 1787, stopped April 1789 3a. Cylinder to Poldice No. 2 engine 3b. Rest of engine to Herland (see 41 below)	Wilson papers
27. Wheal Virgin	Started 1786	20″	Double engine 1. Erected 100 fathoms underground 2. To Wheal Jewel, started 1792, working May 1798	Wilson papers
28. Prince George	Started 1st February 1786	20″	Double engine 1. Stopped 30th October 1786 2. To Seal Hole, started 1st June 1791	Wilson papers
29. Polgooth	Started February 1786	58″	To Cockshead	Wilson papers
30. North Downs (Briggan)	Started February 1786	45″	Double engine	Wilson papers
31. Wheal Mount	Started February 1786	20″	Double engine 1. Stopped 17th September 1787 2. To Mr Gullet, Beerferris mine June 1788 and started that year, burnt down at end of 1791 3. To Mr Crayshaw, Bereferris, started August 1797	Wilson papers
32. Wheal Reeth / Wreeth	Started 8th March 1786	20″	Double engine 1. Stopped December 1790 2. To Wheal Neptune, started July 1791, stopped 30th November 1791 3. To Wheal Clowance (engine owned by Boulton & Watt) 4. To Wheal Jewel, stopped January 1797	Wilson papers
33. North Downs (Wheal Messa)	Started March 1786	42″	To Briggan	Wilson papers
34. Hallamaning	Started 1787	60″	1. To Wheal Ramoth, started December 1796 2. To Herland, erecting May 1798	Wilson papers
35. Wheal Crenver	Started February 1787	60″	In same house as Wheal Jewel 36″	Wilson papers
36. Poldice	Started April 1787	24″	Double engine To Wheal Gorland, started March 1796	Wilson papers

37. Poldice	Started April 1787	58″	Double engine "Working on account of Wheal Unity, since January 1797"	Wilson papers
38. Poldory	Started June 1787	45″	Double engine Working at Ale & Cakes from its first erection	Wilson papers
39. Wheal Maid	Started 1st April 1788	63″	Double engine	Wilson papers
40. Wheal Butson	Started October 1791.	36″	1. Stopped Spring 1793 2. To Bog Mine (near Marazion), started 14th June 1791, working 1794 3. To Wheal Jewel, started November 1796	Wilson papers
41. Herland	Started June 1792	64″	New cylinder and remains of Poldice No. 3 engine (see 9 above)	Wilson papers
42. Cooks Kitchen	May 1793	36″		Wilson papers
43. Cardrew Downs	Started June 1794 ("bought at Nottingham after being burnt there by accident")	20″	Double engine, rotative Stopped January 1796	Wilson papers
44. Hewas	July 1794	45″	Double engine Stopped January 1798; to Treskow, started April 1798	Tann J., 1986; Wilson papers
45. Hallenbeagle	December 1796	52″	Double engine	Wilson papers
46. West Wheal Jewel	1798			Tann J., 1986
47. New Roskear	1798			Tann J., 1986
48. Herland	1798			Tann J., 1986
49. United Mines	1801		6 HP engine and drawing apparatus	Dickinson H. W. & Jenkins R., 1927; Tann J., 1986

Chapter 14
Hornblower and Winwood's double cylinder engines.

Jabez Carter Hornblower, writing in *Gregory's Treatise of mechanics* of 1806, records that his brother Jonathan Hornblower (II) (1752 – 1815) of Penryn had started developing an engine as early as 1776 "if not before".[1] Jonathan patented his engine on 13th July 1781, giving his design fourteen years protection. The fundamental principle behind his design was that it utilised two cylinders, the steam passing from the smaller to the larger cylinder in succession, the steam working expansively. Hornblower's design was the first compound engine; a design that would be perfected by a later generation of Cornish engineers, most notably Woolf. In 1788 Hornblower described the operation of his engines thus:

> "..... ours consists in causing the steam, after it has operated in one cylinder, from thence to pass into another; and there (while it is succeeded in the first by a fresh supply from the boiler) produces a second effect. By this means the steam is employed to the greatest advantage that is possible; for while it actuates the first piston, it retains its full elastic force, but when it is applied to press on the second piston, the communication with the boiler is cut off; and it is then permitted to expand itself, by being admitted into a cylinder of greater capacity."[2]

Jonathan entered into partnership with John Winwood, a Bristol iron founder, Hornblower holding 3/5ths whilst Winwood held 2/5ths.[3]

By 1782 Jonathan had erected his first engine to his patent at Radstock on the Somerset coal field.[4] This engine had cylinders of nineteen and twenty four inches. Boulton and Watt felt that this engine was an infringement of Watt's patent. Although they did not prosecute on this occasion, they did place an advertisement in the Bristol newspapers outlining six features invented by Watt stating that they would prosecute anyone who infringed Watt's patents.[5] This appears to have been a favourite strategy of Boulton and Watt who tended to prefer making threats to taking legal action, possibly because they had doubts as to the robustness of their engine patents and did not want to test them in court.

The first Hornblower double cylinder engine in Cornwall was erected at Penryn, work starting in 1784, and completed in 1787. The engine was built to half scale with

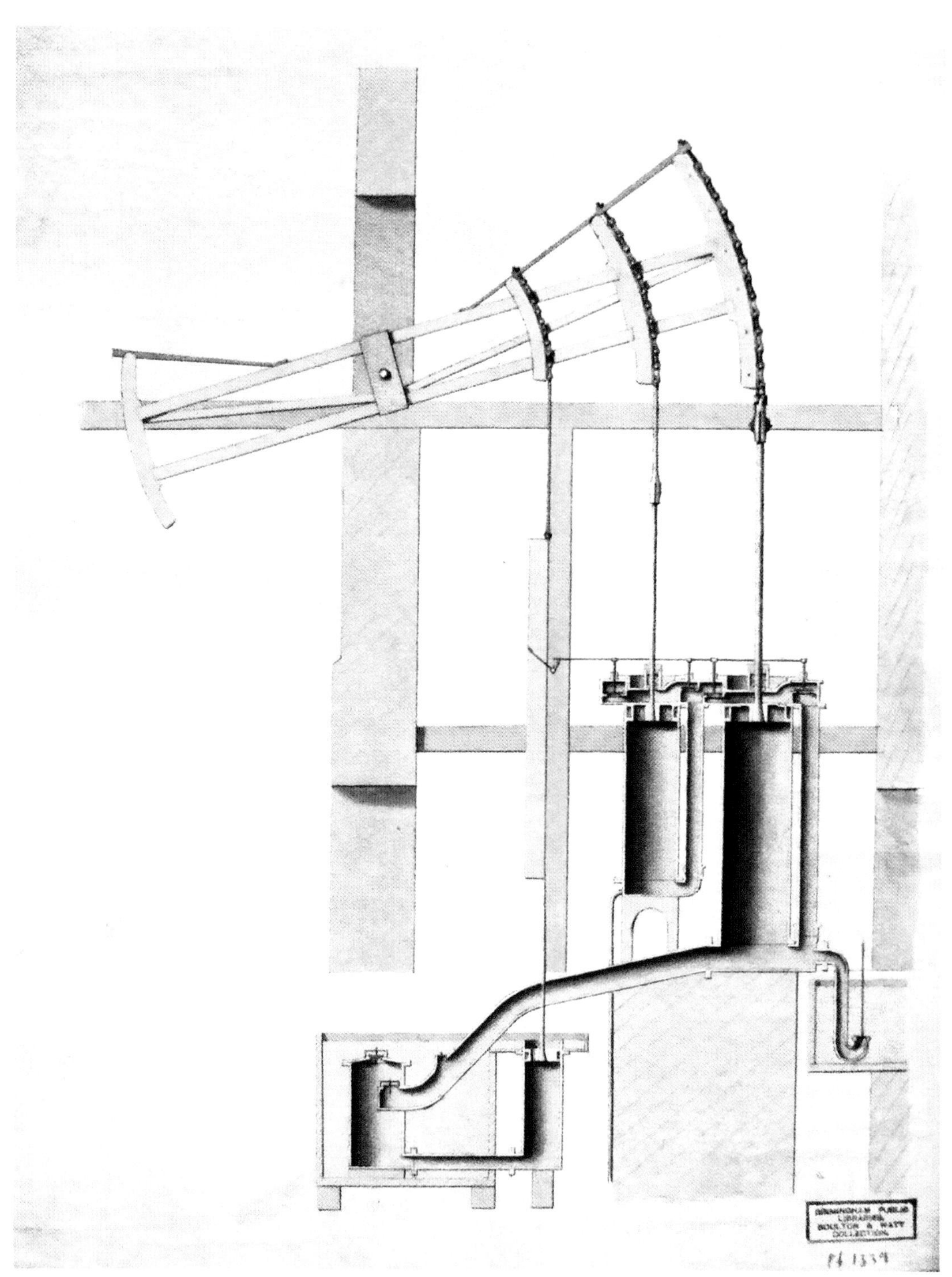

Drawing from the Boulton and Watt collection of Hornblower and Winwood's Radstock engine. This would have been prepared at Boulton and Watt's instigation in regard to Hornblower and Winwood's alleged breach of Watt's patent. Reproduced with the permission of the Library of Birmingham, MS 3147/1339b.

Jonathan Hornblower (1753-1815), arguably the best Cornish engineer of his generation who, but for the times he lived in, could have achieved greatness.

nine and eleven inch cylinders and was probably used to demonstrate Hornblower's design to Cornish adventurers.[6] The completion of the Penryn engine coincides with the publication by Hornblower and his partner John Winwood of "*An address to the Lords, Adventurers, and others concerned in the mines of Cornwall*". This eight page document, dated 1st May 1788, outlines the principles of Hornblower's engine and how it differed from Watt's patents. The document concludes:

> "We are ready to treat with any set of adventurers, and undertake either to alter any of their present engines, or build new ones on our own principles, allowing them terms so much to their advantage, as they cannot reasonably object to. And also to indemnify them by a proper bond, against all costs and damages at

law that may ensue from a prosecution by any other patentees.

Having now, gentlemen, said all that we think is necessary; to convince you, not only of the superiority of our invention, but that our principles and construction are so very different, as not to infringe on any others; it remains with you to seriously consider, how far it deserves your patronage and support, or whether you will suffer yourselves any longer to be deprived of the use of so great an invention, by the misrepresentation, influence, and threats of interested persons, who, as we have clearly proved, have neither law nor equity on their side."[7]

In spite of this marketing campaign, orders for the double cylinder engine were initially slow in arriving, due in no small part to the depressed state of mining in Cornwall. In a letter dated 12th March 1789 Jonathan, writing to his uncle Josiah Hornblower (1729 – 1809) in America, comments:

"We have hitherto had no success as the copper trade is so low, and tin too as to entirely check the spirit of mining. Therefore Watt's engine's are still standing (a few of them) but the greater part are stopt........"[8]

In such a stagnant market it comes as no surprise that Boulton and Watt felt threatened by Hornblower and Winwood's attempts to establish themselves in Cornwall. By the beginning of 1790 Boulton and Watt were considering what steps they could take to counter the threat posed to their Cornish monopoly by Hornblower's engine. For example in late 1789 /early 1790 Boulton and Watt were contemplating buying out Winwood's share in the partnership to neutralise the threat, however, this particular strategy did not come to fruition.[9]

Letters dated 28th and 30th January 1790 show that James Watt and Thomas Wilson were colluding to buy shares in Tincroft mine so that Wilson could attend Tincroft meetings.[10] The reason for Watt and Wilson's interest in Tincroft was the mine's intention to employ one of Hornblower and Winwood's engines and, in Watt's words: "Without evidence we can do nothing".[11]

On 14th February 1791 Hornblower and Winwood started an engine with 21 inch and 27 inch cylinders on Tincroft Mine,[12] the event being recorded in the pages of the *Sherborne Mercury* of the 28th February 1791:

"On Monday the 14th instant was set to work on Tin Croft Mine, in the parish of Illogan in Cornwall, Messrs. Hornblower and Winwood's new – invented steam – engine, an event at which every person interested will have reason to rejoice. This powerful machine, which is infinitely superior to everything of the kind hitherto known, must by lessening the expense of mining raise the spirit of adventure. It seems to be the ne plus ultra of mechanical invention, and does equal honour to the ingenuity of the inventor, and to the liberality and public – spirited patronage of the Tincroft adventurers."[13]

Hornblower and Winwood were not slow to sing the praises of the Tincroft engine; an advertisement of 30th July 1791 reading:

"Messrs. Hornblower and Winwood have, by their Engine at Tin Croft, exhibited a Machine which evidently surpasses every other of the kind in a double proportion."[14]

In 1792 Richard Trevithick Senior carried out a twenty four hour trial of this engine during which time he recorded a very respectable duty of 16,609,672.[15]

When the Tincroft engine was started Boulton and Watt appear to have seriously contemplated a prosecution of Hornblower and Winwood for infringement of Watt's patents. On the 9th February 1791, in a letter to Wilson, Watt wrote:

"In respect to Tin Croft, though we shall go on slowly & with due deliberation yet we shall certainly proceed against them in such way as shall most effectually operate against them. We depend on you & Mr Murdock to set about procuring proper & if possible respectable evidence 1st that they have used the power of steam & not the pressure of the atmosphere to act upon the piston of the engine, 2nd that they draw out the air by means of an air pump, 3rd that they do not condense the steam in the steam vessel itself, where the powers of steam are exerted to work the engine, but in some place separated from it, 4th that they use Grease to make the piston air tight.

Until distinct and clear evidence can be procured on these heads we can do nothing in the way of Law, you will therefore lose no time in doing the needful, so long as the profits of the engine business will pay the lawyers so long shall We defend our case."[16]

By the 18th March 1791 Boulton and Watt had been granted permission by Jonathan Hornblower (II) to examine the Tincroft engine; William Murdock and Edward Rogers being deputed to carry out the inspection.[17] However it appears that Boulton and Watt had not approached the Tincroft adventurers with the request and, in consequence they withheld permission. The situation seems to have been compounded by erroneous reports, apparently spread by Boulton and Watt's Cornish agents, that "with the authority of the Lord Chancellor the engine was to be taken to pieces for examination".[18] The situation continued into the summer of 1791, on 13th August Watt wrote:

"...... these Gentlemen (*Hornblower and Winwood*) did indeed consent that B & Ws agents should examine the Engine but on their arrival there at the appointed time, were refused a particular examination of the contested parts by the Capt (ain)s on the mine & that after repeated applications to the adv(enture) rs to the same purpose no consent of their part has been obtained."[19]

By early September 1791 Boulton and Watt had taken legal opinion regarding the situation and were advised by their counsel to "if possible avoid going to law".[20] Boulton and Watt appear to have followed this advice as they did not proceed with regard to Hornblower and Winwood's engine at Tincroft. That is not to say that they abandoned their campaign against Hornblower. In the spring of 1792 Jonathan Hornblower (II) applied to Parliament for an extension to his engine patent of 1781, which was due to expire on 13th July 1795.[21] As was to be expected there was extensive lobbying both for and against the patent extension. Not surprisingly Boulton and Watt did everything in their power to ensure that Hornblower did not get his hoped for extension for they felt that this reflected a much wider issue; Matthew Boulton commenting on 22nd March 1792: "remember that it is Cornwall ver(su) s B(oulton) and W(att)".[22] Wilson was urged to supply whatever evidence he could muster, Boulton advising Wilson not "to spare a few Guineas in bribing men to be honest".[23] Unfortunately for Hornblower Boulton & Watt's influence prevailed; Hornblower's application being defeated by 63 to 28 on the bill's second reading, although it is worth noting that the majority of Cornish members voted in favour of Hornblower's application.[24]

In the summer of 1792 Thomas Wilson, at the behest of Boulton and Watt published as pamphlet entitled "*A comparative statement of the effects of Messrs. Boulton and Watt's steam engines with Newcomen's and Mr Hornblower's*" which resulted in a war of words fought in the pages of various regional newspapers, albeit with little material impact on either side.

In spite of his failure to gain the extension of his patent and the ongoing war of words Hornblower remained sanguine in regard to Boulton & Watt. The success of the Tincroft engine led to a number or orders from Cornish mines. By late 1792 Hornblower's star was very much in the ascendant in Cornwall and threatened to eclipse that of Boulton and Watt. On 15th December 1792 Hornblower commented to his friend Davies Gilbert (1767 – 1839)

"I am, however, pretty easy, for they build no engine and we now have nine in hand."[25]

The success in securing orders for his engine is confirmed in a letter written by Jonathan to Josiah Hornblower on 28th April 1793:

"We have two engines at work in this county and 8 more erecting some of them very large......"[26]

The two engines already at work were the Tincroft engine, already discussed and an engine with fifteen and nineteen inch cylinders erected at Swanpool Mine near Falmouth which was started in September 1792. Of the eight engines under way in April 1783 five were being erected by Jonathan and three by his brother Jethro Hornblower (1746 – 1820).[27]

Like Boulton and Watt, Hornblower and Winwood charged mines a premium for the use of their engines based on fuel saving. John Winwood outlined the formula for calculating the premium's in a letter to the Wheal Pool adventurers dated 27th February 1796:

> "The difference of the power of Our Engine compared with Mr Watt's we find to be as thirty is to eighteen, and therefore we expect to be paid for such proportional share of the value of what Coal you might have consumed from the setting to work of the Engine, to the expiration of the term of our patent, which was on the 13th July 1795".[28]

Table 4. List of Hornblower's engines erected in Cornwall.

Mine	Date started	Cylinders	Notes	References
1. Tincroft	February 1791	21″ & 27″	Re-built as a whim 1. Stopped 1794 2. To Wheal Fortune (Illogan)	Barton D. B., 1966; Farey J., 1827; Tann J., 1981; Wilson T, 1792; Wilson papers.
2. Swanpool	September 1792	15″ & 19″	Re-built as a whim 1. Stopped 1793. 2. To Wheal Crenver 1798	Barton D. B., 1966;, Tann J., 1981; Tann J., 1996; Wilson papers.
3. Baldhu	September 1793	21″ & 27″	1. Stopped 1795. 2. To Wheal Towan 1799	Barton D. B., 1966; Pole W., 1844; Tann J., 1981; Tann J., 1996; Wilson papers.
4. Tresavean	September 1793	30″ & 36″		Barton D. B., 1966; Pole W., 1844; Tann J., 1981; Tann J., 1996; Wilson papers.
5. Lostwithiel	September 1793	15″ & 19″	1. Stopped 1794 2. To Wheal Tregoth-nan 1796-7 3. To East Pell, St. Agnes 1798-9	Barton D. B., 1966; Pole W., 1844; Tann J., 1981; Tann J., 1996; Wilson papers.
6. Wheal Pool	October 1793 (Tann J., 1981 states 1794)	30″ & 36″	1. Stopped 1795 2. To Wheal Provi-dence 1797	Barton D. B., 1966; Pole W., 1844; Tann J., 1981; Tann J., 1996; Wilson papers.

7. Wherry Mine	November 1793	21″ & 27″	Stopped 1797	Barton D. B., 1966; Pole W., 1844; Tann J., 1981; Tann J., 1996; Wilson papers.
8. Wheal Margaret	3rd November 1793	21″ & 27″		Barton D. B., 1966; Pole W., 1844; Tann J., 1981; Tann J., 1996; Wilson papers
9. Wheal Unity	December 1793	45″ & 53″		Barton D. B., 1966; Pole W., 1844; Tann J., 1981; Tann J., 1996; Wilson papers.
10. Tincroft No. 2	January 1794	45″ & 53″		Barton D. B., 1966; Tann J., 1981; Tann J., 1996; Wilson papers.

Of particular interest due to its peculiar situation is the Hornblower engine erected at the Wherry Mine at Penzance, arguably one of the most extraordinary mines in Cornwall. In 1843 J. Y. Watson describes the mine thus:

> "The Wherry Mine near Mount's Bay, Penzance, as a submarine mine, is celebrated as the most extraordinary in the history or Cornwall; and was established on a shoal 720 feet from the beach at high water; the rock is covered about ten months in the twelve, and the depth of water on it at spring tides 19 feet; and in winter the sea bursts over the rock in such a manner as to render all attempts to carry on mining operations unavailing; veins of tin were first discovered in an elvan course running in front of Penzance in the early part of the last century, and attempts were made to work it, but abandoned as hopeless. Notwithstanding all the difficulties, however, a poor miner of Breage, named Thomas Curtis, in 1778, had the boldness to renew the attempt, and after innumerable difficulties succeeded in forming a water - tight case, as an upper part of the shaft, against which the sea broke, while a communication with the shore was established by means of a wooden frame bridge.
>
> As the work could be prosecuted only during the short period of time when the rock appeared above water, three summers were consumed in sinking the pump shaft; and the use of machinery becoming practicable, the water tight case was carried up a sufficient height above the highest spring tides..... In the autumn of 1791, the depth of the pump shaft, and of the workings was 26 feet: 12 men were employed for two hours at the wins (*author's note: winch*) for hauling the water, while six men were teaming from the bottoms into the pump...... After a time a steam engine was erected on the green opposite, and

hanging rods from it, carried along the wooden bridge to the mine......"[29]

The choice of steam engine was something of a contentious issue, the matter being discussed at Adventurers meetings in August and September 1792. Thomas Wilson offered to "build & set to work" a Boulton and Watt engine at a cost of £550. However at a meeting held on the 16th October 1792, on the recommendation of Mr Moyle "a particular friend of Hornblowers", the meeting agreed to erect a Hornblower engine if Hornblower and Winwood would agree to erect it at the price offered by Wilson. An agreement was duly signed and work started on the engine.[30] The Wherry adventurers decision to choose a Hornblower engine, rather than a Boulton and Watt, infuriated James Watt who wrote to Wilson on the 23rd October 1792 expressing his opinion in rather bitter terms:

> "The Wherry affair is most provoking.........& if the engine causes loss to the mine you will have your revenge, by reproaching the adv(enture)rs who negotiated this wise affair. In regard to the proposal of erecting an engine to compare with it I am against it, because we know not what new Jimbol they have got, or how much more (of) our engine they may Steal, & because our trial with Bull will probably be over if we lose that it will answer no end to confute Hornblower, & if we gain it we shall probably bring the other over the Irons next term........ I am really sorry for you that you should meet with such mortifications in our cause but we cannot remedy them & best is to be patient & quiet at present, perhaps you may have it in your power to piss upon them in your turn. By no means sell out of a good mine but when opportunity offers stick to their ribs & give them no advice.[31]

The chemist Charles Hatchett (1765 – 1847) visited the Wherry Mine on 13th May 1796 observing:

> "The shaft of the mine is in the sea at about 70 fathoms distant from the House which contains the Steam Engine ..... There is a wooden platform which is supported by long posts which reaches from the Engine House to the shaft. The Engine is worked by Steam and is on Hornblower's principle with two cylinders, one of which receives the superabundant steam of the other, and conducts it into a condensing vessel. The steam engine works a horizontal rod which at the end over the shaft moves a Road and Bucket."[32]

As we have seen, orders for his engines were coming in but Jonathan's failure to secure an extension to his patent led him to consider emigrating to America on its expiry. In considering this course of action, Jonathan was no doubt influenced by his uncle Josiah who went to America in 1753 to erect an engine. Josiah's voyage was so unpleasant that he vowed never to cross the Atlantic again and remained in

The remarkable Wherry Mine at Penzance, an engraving from the Transactions of the Royal Geological Society of Cornwall, 1822. At the far right can be seen an angle-bob which converts the to-and-fro motion of the engine along the platform to up-and-down in the shaft.

America.[33] On 28th April 1793 Jonathan wrote the following to his uncle Josiah:

> "..... if we can not gain a longer time we shall be little benifitted for we have but three years more to expire of our patent. I wish to know how the climate of New York wou'd be likely to agree with me. The cold I think I shou'd not mind so much as the heat; you have a certain fly too which is very troublsom. If I were to go over what prospect wou'd there be of getting into immediate livelyhood? I shou'd not like to quit a certainty of something and go in quest of fortune at this time of life (40) without good prospects. Pray do you burn wood or coals at your engine? And what is the price of provisions particularly Bu(t)chers Meat in your part of America?"[34]

When the 1781 patent did expire Jonathan Hornblower chose to remain in Cornwall rather than emigrating, turning his attention to developing a rotary engine. Jonathan patented this engine in 1798 and by 1st June 1799 he was able to write to Josiah informing him that six of his rotary engines were in course of erection.[35] Unfortunately due to its complexity the engine, was not a success, John Farey, in his *Treatise on the steam engine* of 1827 observes:

> "Mr Hornblower made many attempts to bring this ingenious project to bear, but the engine could not be made to answer, on account of the friction and leakage of the packing, which is more than twice as much as is required in a common engine exerting the same force........ The stopping and starting of the pistons was always attended with a violent jerk, and the clamps to prevent the retrograde motion, were very subject to wear out of order, and then to break

or run back."[36]

Jonathan did not completely abandon the compound pumping engine after the expiry of his patent. In June 1799 he said that he was "erecting two very large engines of the kind for which I obtained my first patent". The new engines were certainly much larger than of his previous pumping engines having sixty and fifty three inch cylinders. One of these engines was erected on Wheal Unity.[37]

The success of Boulton and Watt's action against Jabez Carter Hornblower and J. A. Maberley in 1796, confirmed at appeal in January 1799, gave Boulton and Watt the confidence in their patents to start actively chasing outstanding premiums. Until this time Boulton and Watt had never taken any legal action against Jonathan Hornblower (II) for infringement of their patents.[38] However on 16th October 1799 Messrs Weston, acting on Behalf of Boulton and Watt sent the following letter to Thomas Wilson:

> "It being determined when Mr Robinson Boulton was last in Town to bring an Action against Jonathan Hornblower to establish the fact of infringement on Mr Watt's Patent, in the Engines erected by him in Partnership with Mr Winwood, in hopes that a verdict against him may induce all the mines where his Engines are erected, to agree to Boulton and watt's demands, we inclose you a Writ against him, which we request you will get Mr. Edwards to serve with as little delay as possible, in order that the Action may be tried at the Sittings after next Term."[39]

In addition to the writ against Hornblower, writs were also "sued" "against the Adv(venture)rs in all Mines having Hornblowers Engines".[40] The implication was that should the case against Hornblower go in favour of Boulton & Watt, they would then take action in law against the mines which had employed Hornblower's engines, to recover outstanding premiums.

On the 31st January 1800 James Watt Junior wrote to Thomas Wilson informing him that he hoped that the case against Jonathan Hornblower (II) would be "tried in the course of the next month" and that William Murdock had made a "capital model of Horners engine" to support their case.[41] On the 11th February 1800 it was noted that Hornblower had managed to delay proceedings "until the Sittings after next term".[42] It was the opinion of Weston that the Hornblower case would not be heard before mid July 1800.[43] At the end of May the possibility of an alternative solution began to emerge. This seems to have originated with John Vivian, acting as an informal spokesman for various adventurers who would be affected should legal action go against Jonathan Hornblower. On 30th May 1800 John Vivian wrote to Boulton and Watt suggesting an alternative to legal action:

> "I am sorry to find that you are still engaged in Law Suits on Account of

your Engines especially as I fear I shall myself be interested in the Event – Would it not be more prudent for both parties to endeavour to agree upon some compromise than spend such sums of Money in Law Suits will probably leave but little Advantage to the gainers."[44]

In reply to Vivian's approach Boulton & Watt replied on the 3rd June 1800 that they would prefer "to terminate amicably our disputes in Cornwall rather than be recurring to Litigati(o)n". However it was clear from their letter that any proposal should come from the adventurers concerned rather than from Boulton and Watt.[45] Negotiations appear to have been successful. On 21st June 1800 Robinson Boulton (on behalf of Boulton & Watt) wrote:

"We have agreed to postpone the trial with Hornblower till the Sitting after Michaelm(a)s Term in consequence of Mr Vivian's representation of the amicable dispositions which pervade the majority of parties interested."[46]

This amounted to a mutual and pragmatic settlement between the adventurers and Boulton and Watt, both of whom would avoid the cost and inconvenience of expensive legal action. It was also a tacit acceptance on the part of the adventurers that Hornblower and Winwood had infringed Watt's patents; which, given their rather jaundiced views of Cornish adventurers, must have caused Boulton and Watt a certain degree of satisfaction. Boulton and Watt's decision was also no doubt strongly influenced by the expiry of their patent. If Boulton and Watt had been able to continue claiming premiums beyond 1800 they might not have been so accommodating. Thus by September 1800 adventurers at mines which had used Hornblower and Winwood's engines were being requested to pay a premium to Boulton and Watt. After negotiation and in some cases arguments regarding the amounts owed, patent dues were eventually paid to Boulton and Watt.[47] Pole's *Treatise on the Cornish pumping engine* contains the following list:

"List of Engines on Hornblower's construction erected in Cornwall; together with the amount of patent dues demanded for each by Messrs Boulton and Watt after the result of the trial.

	£	s	d
Tincroft	2569	7	0
Wheal Unity	2100	0	0
Tresavean	1000	0	0
Wheal Margaret	*548	8	0
Wherry	*432	2	0
Wheal Pool	150	0	0

Wheal Providence	124	16	0
Baldice	120	16	0
Wheal Tregothnan	32	2	0
East Pell Adventurers	94	3	0
Lostwithiel	14	0	0
Wheal Towan	40	0	0
Total	7225	14	0

To the two amounts marked * the words "not yet paid are appended: therefore we may infer that not all the money had been received by the patentees."[48]

The settlements negotiated with the various adventurers seem to have satisfied Boulton and Watt who chose not to proceed with their action against Jonathan Hornblower (II). This is confirmed (although differing in details) by Boulton and Watt's counsel Mr Weston who in 1808 commented in regard to the lack of legal action against Hornblower:

> "I do not find there was any judgement by default or otherwise. I believe the fact was that Mr Robinson Boulton went into Cornwall and frightened the users of Jonathan Hornblower's engines out of about £40,000 which with being content, the action against Jonathan was dropped of course."[49]

From a historian's point of view the lack of a legal confrontation between Boulton and Watt and Hornblower is disappointing. It is far from certain that Hornblower and Winwood's engine was an infringement of Watt's patents and it would have been interesting to have seen the merits of both cases tested in a court of law.

William Pole sums up Jonathan Hornblower's contribution to the development of the steam engine thus:

> "His invention of the double – cylinder engine would alone, if he had done nothing else, cause his name to be remembered as long as the steam engine exists, or its history remains on record."[50]

It is unfortunate that Jonathan Hornblower's work on pumping engines should have coincided with the period of Watt's patent. Hornblower was one of the great Cornish engineers and it is a tragedy that his contribution to the development of the Cornish pumping engine was impeded by the straitjacket of the Watt patent. If Jonathan's patent had been extended beyond 1795, as it might well have been in the absence of Boulton and Watt's intervention one wonders what Hornblower might have achieved. As early as 1791 Davies Gilbert was suggesting to Hornblower that he "work with steam considerably stronger than the Atmosphere......."[51] If Hornblower had followed Gilbert's advice he would have significantly pre-empted Arthur Woolf who very successfully applied high pressure steam to Hornblower's double cylinder engine in the early nineteenth century. However it would require the use of Richard Trevithick Junior's cylindrical boilers to produce steam at these elevated pressures; these would arrive eight years later. Whether, given a clear field, Hornblower would (or indeed could) have pioneered the use of "strong steam" is a moot point but interesting, if ultimately unprofitable, speculation on what might have been.

Chapter 14 References

1. Gregory O., 1806
2. Hornblower J. & Winwood J., 1788 cited in Torrens H. S., 1982
3. Harris T. R., 1976; Wilson papers: AD1583/4/54
4. Dickinson H. W. & Rhys Jenkins 1927; Harris T. R., 1966
5. Dickinson H. W. & Rhys Jenkins 1927
6. Dickinson H. W. & Rhys Jenkins 1927; Torrens H. S., 1982
7. Hornblower J. & Winwood J., 1788, cited in Torrens H. S., 1982
8. Cited in Torrens H. S., 1984
9. Wilson papers: AD1583/1/5 & AD1583/4/8
10. AD1583/4/4, AD1583/4/5
11. Wilson papers: AD1583/4/4
12. Barton D. B., 1966; Farey J., 1827, Wilson T., 1792
13. Cited in Barton D. B., 1966
14. Cited in Wilson T., 1792
15. Wilson T., 1792
16. T. Wilson Papers: AD1583/4/47
17. T. Wilson papers: AD1583/4/54
18. T. Wilson papers: AD1583/4/64
19. T. Wilson papers: AD1583/4/74
20. T. Wilson papers: AD1583/4/ 76
21. Wilson papers AD1583/9/11
22. Wilson papers: AD1583/5/18
23. Wilson Papers: AD1583/5/16
24. Farey J., 1827; Harris T. R., 1976, Tann. J., 1981
25. Cited in Todd A. C, 1961
26. Cited in Torrens H. S., 1984
27. Barton D. B., 1966; Harris T. R., 1976; Tann J., 1981; Torrens H. S., 1984
28. Wilson papers: AD1583/9/11
29. Watson J. Y., 1843
30. Wilson papers: AD1583/11/79
31. Wilson papers AD1583/5/54; Joseph, P., 2012
32. Cited in Harris T. R., 1976
33. Allen J. S. & Rolt L. T. C., 1997
34. Cited in Torrens H. S., 1984

35. Farey J. 1827; Torrens H. S., 1984
36. Farey J., 1827
37. Torrens H. S., 1984
38. Torrens H. S., 1984
39. Wilson papers: AD1583/11/19
40. Wilson papers: X208/39
41. Wilson papers: X208/8
42. Wilson papers X208/10
43. Wilson Papers X208/18
44. Wilson papers X208/22
45. Wilson papers X208/22
46. Wilson papers: X208/24
47. Wilson papers: AD1583/11/40, 41, 42, 43, 45, 46, 47, 50, 52, 54
48. Pole W., 1844
49. Cited in Tann J., 1981
50. Pole W., 1844
51. Cited in Todd A. C., 1961

Chapter 15
The engines of Edward Bull.

Whilst there are significant doubts as to whether Hornblower and Winwood's double cylinder engine was an infringement of Boulton and Watt's patents there can be no such doubt regarding the engines erected by Edward Bull (c. 1759 – 1798).

The earliest reference to Edward Bull is as an engine man at Bedworth Colliery near Coventry where in 1779 he was employed at eleven shillings a week. In 1781 he was on the staff of Boulton and Watt and had been sent to Cornwall as an engine erector.[1] By November 1788 he was employed at Wheal Virgin, having presumably left Boulton and Watt's employment. Bull's appointment appears to have been as part of an agreement between Boulton and Watt and Wheal Virgin. With regard to this post James Watt commented on 26th November 1788:

> "Bull should also be informed that it was by our desire he was employed at Wheel Virgin & that by the agreement we can displace him whenever we find it necessary."[2]

It seems rather strange that Boulton and Watt should have the authority to "displace" Bull who was employed by Wheal Virgin, however right to veto engine men appears to have been a typical condition of their agreements with Cornish mines.

By September 1789 Bull was carrying out alterations, unauthorised by Boulton and Watt, to the Wheal Virgin engines. The nature of Bull's alterations are somewhat obscure although they appear to have involved coupling the cylinders of the 50-inch West Wheal Maid engine and the 58-inch Wheal Virgin Elvan engine together. This attracted the attention and disapproval of Jonathan Hornblower (II), although no legal action resulted.[3] Bull's experiment with the Wheal Virgin engines does not appear to have been a success; on 10th November 1789 Matthew Boulton commented:

> "Although we did not expect much good from the alteration of the engines at Wheal Virgin, yet nevertheless we did not expect it to be so bad as they are said to be – We never advised makeing the Alteration, but now it is done, We do advise (before the expence be decidedly thrown away) that Mr Murdock be employed to examine it......."[4]

A typical Bull engine. Most notable is the inverted cylinder and the lack of a main beam (which would reduce build cost). In the absence of a main beam the plug rod and hence the working gear was driven from the balance bob.

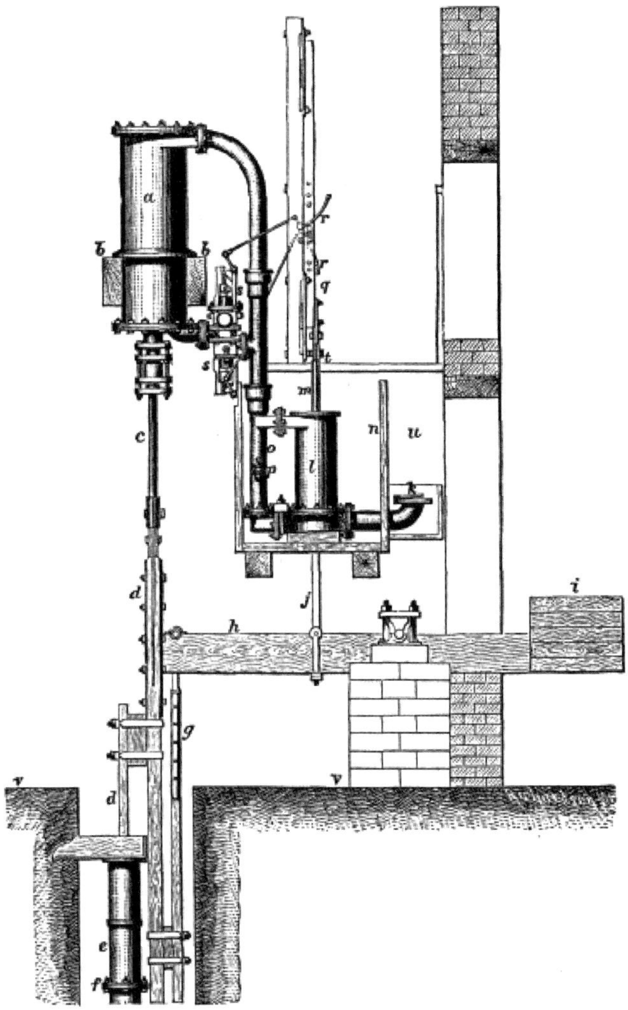

Whilst Bull's modifications to the Wheal Virgin engines were not a success, it did establish him as one prepared to challenge the absolutism of Boulton and Watt. This would prove attractive to Cornish adventurers shackled, as they saw it, by Boulton and Watt's premiums. Certainly by August 1791 Bull appears to have been acting as an independent engineer/erector.[5] In August 1791 "Ned" Bull had agreed to build a small engine at Wheal Rose, a small lead mine near Helston. Bull agreed to:

> "build the Engine, furnish pumps & every article & keep them in repair & to work it to the depth of 45 fathoms for 42s per month to encrease to 70 in proportion to the water".[6]

It is doubtful as to whether this engine was ever built. In September 1791 Bull approached Boulton and Watt with a view to erecting engines to their patent, albeit without reference to either Thomas Watson or William Murdock.[7] It comes as no surprise that Bull's approach was rejected, Watt writing to Wilson on 10th October 1791:

> "In respect to Bull the less we have to do with him the better if he applies to you on our terms & brings respectable persons as principals you will fix the premium with him & take his order for the size of the Engine but we will not be directed how to make it.
>
> Had we agreed to have let him made one of our Engines in such a manner as

he pleased, he would have made a bad thing & we should have had our share of the disgrace as it now stands."[8]

Boulton and Watt's refusal of Bull's request meant that, should Bull continue to erect engines, he would find himself in direct opposition to his former employers. Indeed within a year Bull would find himself in conflict with Boulton and Watt. By October 1792 Bull had erected an engine at Balcoath (referred to in the Wilson papers as Godolphin). On 14th October 1792 Watt wrote:

"..... as to Godolphin vizt that you should inform the adventurers that whatever additions they make to the Engine, on our principles, we shall charge them for, whoever makes them, That Bull's Engine is entirely on our principles & that we are now bringing an action against him which will be tried next term."[9]

To facilitate legal proceedings William Murdock produced drawings of Bull's Balcoath engine.[10] Murdock's drawings of the engine have survived, one of which is reproduced on page 139 and this may be seen as typical of the Bull design. The most obvious difference between Bull's engine and those erected by Boulton and Watt was Bull's use of an inverted cylinder, sitting directly over the shaft. The piston connecting rod exited the cylinder through a gland at the bottom of the cylinder and connected directly to the pump rod in the shaft without the need for a main bob / beam. Dispensing with the bob would reduce the first cost of the engine, an attractive proposition for cash strapped adventurers. However innovative the inverted cylinder was, the rest of the engine was, demonstrably, a Boulton and Watt engine with separate condenser and, as such an infringement of Boulton and Watt's patents.

There is some evidence to suggest that on occasion Bull attempted to avoid using Watt's condenser and air pump arrangement by working engines "syphonically". Specific details are lacking but probably involved the use of a column of water to draw steam from the cylinder. It is thought the at least four Bull engines, all erected in 1793, worked as "syphonic" engines these included the Wheal Rose 45″, the Carsize Wood 30″, the Wheal Treasure 45″ and the Wheal Treasure 63″.[11] It is probable that Bull revisited the concept of syphonic working on his 1797 Pednandrea engine.

With regard to Boulton and Watt's decision to proceed in law against Edward Bull it was noted in Chapter 12 that Bull became a pawn in a wider game. Elements within the mining interest in Cornwall had long felt that the engine premiums demanded by Boulton and Watt were a heavy yoke to bear. It was also felt that the patent might not stand up to examination in a court of law. If the validity of the patent could be overturned in law the Cornish adventurers would be freed from the engine premiums. Without doubt Boulton and Watt's action against Bull was seen as a test case by many Cornish adventurers some of whom were giving Bull their financial support. Bull was not a rich man and could not have fought what would turn out to be a lengthy litigation against Boulton and Watt without that support. Bull was able to

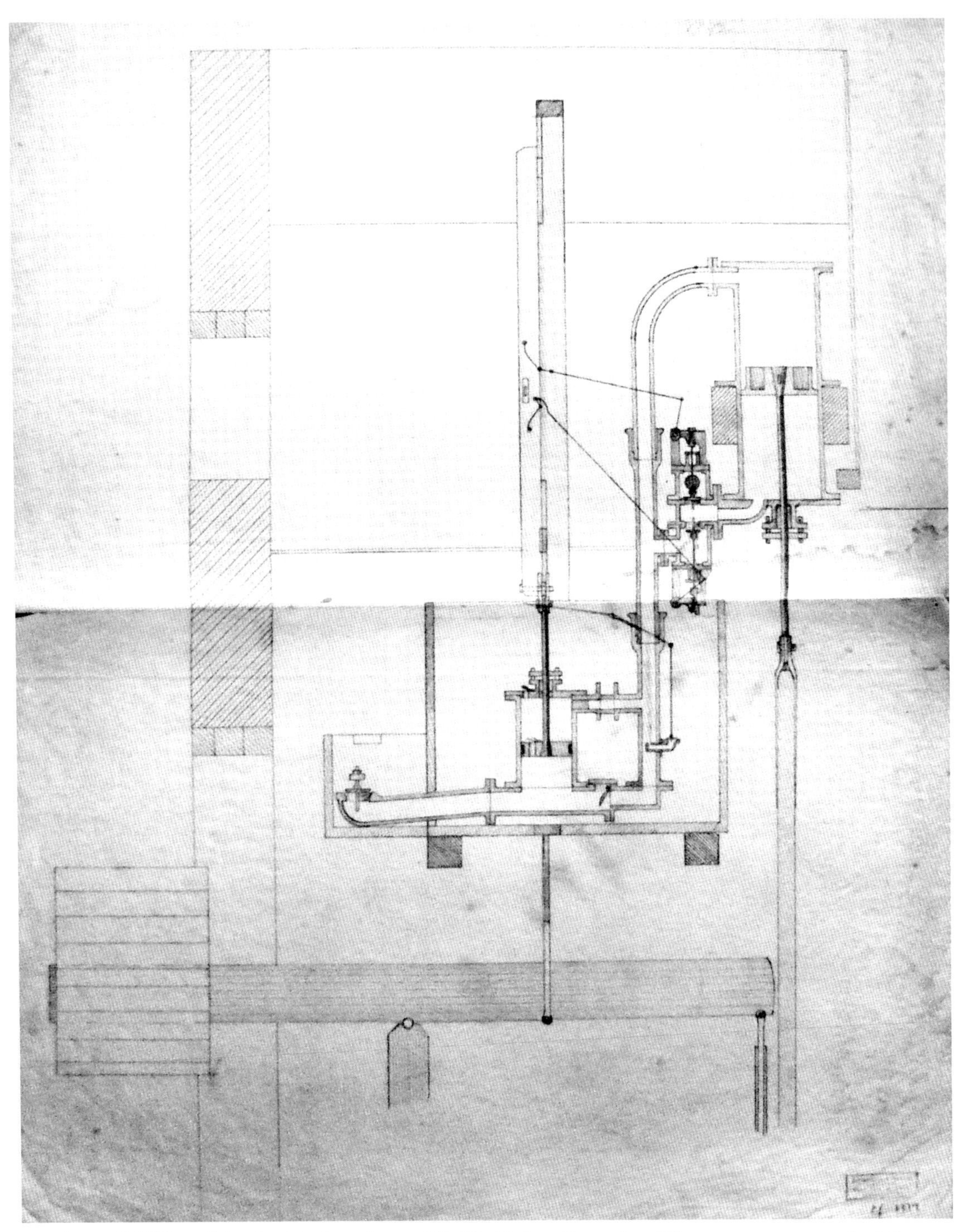

One of William Murdock's drawings of Bull's Balcoath engine. These were prepared at the behest of Boulton and Watt to support their legal actions against Bull. In spite of the innovative looking design, Bull's engines were proved to have breached Watt's patent in a number of ways. Reproduced with the permission of the Library of Birmingham, MS 3147/1339a.

boast that "the county" wanted Boulton and Watt brought to trial and "that it would be no expence to him".[12] There was definitely a great deal of truth behind Bull's assertion. United Mines, for example, were meeting at least some of Bull's costs.[13] That Bull was receiving financial support from Cornish mines was certainly believed by both Matthew Boulton and James Watt.[14] Boulton and Watt were at some pains to prove this as it would mean that, "by showing that improper means have been used to prejudice mens minds against us in Cornwall", they could get the case heard in Middlesex where, in all probability, they would receive a more sympathetic hearing than they would in Cornwall.[15] This apparently was not lost on Bull who is reputed (by Watt) to have stated "that if the cause were tried in Cornwall he should win but not if in Westminster".[16] There was certainly some debate regarding a suitable venue but by mid May 1793 it was decided that the case would be heard in Middlesex.[17]

As previously noted the case of *Boulton and Watt v Bull* was tried in the Court of Common Pleas on 22nd June 1793 in front of Lord Chief Justice Ayre. The court found that Bull had infringed Watt's patent "subject to the opinion of the Court" as to whether the patent was valid in law. To establish whether the patent was valid would require further recourse to law, the case being heard in 1794 and 1795.[18]

Whilst the 1793 decision did not wholly vindicate Boulton and Watt's position it did give them the confidence to pursue mines which had installed Bull's engine. For example on 6th March 1794 Watt wrote to Wilson regarding Bull's engine at Wheal Treasure:

> "Mr Weston says in regard to Wh(eal) Treasure, that we ought not to agree to anything less than paying us the arrears, & either to leave off using Bull's Engine or to pay us for it, the same as if we had made it, & even this he considers a relaxation, as it is freeing them from the damages they would be subject to for infringing our Patent."[19]

The decision also gave Boulton and Watt the confidence to pursue Bull further. On the 22nd March 1794 the Lord Chancellor "granted an Injunction to restrain Mr. Bull from erecting any more engines upon Mr Watt's plan – and from compleating those he had in hand; but the Engines now at work are not at present to be restrained from going on".[20] On the 29th March Bull failed to convince the court to allow him to complete engines he was constructing for Ding Dong and Hallamaning mines.[21] Boulton and Watt made their position very clear:

> "We do not intend by any Means to allow Bull to make any more Engines, besides those in hand; - and now only on proper Terms."[22]

In the case of Hallamaning / Retallack at least Bull appears to have put the engine to work after the 22nd March in contempt of the injunction and to the considerable displeasure of Boulton and Watt who were certainly contemplating legal action.[23]

It is arguable that Bull saw the injunction of 22nd March 1794 as a temporary setback to his engine building career. Assuming that the courts found Watt's patent invalid there would be nothing to stop Bull resuming building engines. The second case (also *Boulton and Watt v. Bull*) was heard on 28th and 29th June 1794 and on the 31st January and 3rd February 1795 although a decision was deferred until 16th May 1795.[24] On the 16th May 1795, in the Court of Common Pleas, Lord Chief Justice Ayre and Mr Justice Rooke found in favour of the patent whilst Mr Justice Heath and Mr Justice Buller found against. As the court was equally divided no judgement was given.[25] This stalemate was not a good decision for Boulton and Watt as the validity of the patent remained ambiguous. However nor was the decision good for Bull. Whilst the Court had not found in favour of Boulton and Watt it had not found in favour of Bull. In consequence Lord Rosslyn refused to dissolve the injunction of 22nd March 1794 on the grounds that whilst there were doubts regarding the validity of the patent, it was a patent of long possession and Lord Rosslyn proceeded on the ground that he would not disturb that possession, the decision being taken on 20th July 1795.[26]

In common with the Hallamaning engine the Bull engine at Ding Dong was also completed after the injunction of 22nd March 1794. Like the Balcoath engine the Ding Dong engine was of the inverted type.[27] Bull appears to have attempted to have tried to circumvent this injunction by using Richard Trevithick Junior to complete the engine.[28] That said, whoever completed the engine did not alter the fact that the engine was, in the eyes of the law, a piracy. In consequence on 11th June 1795 an injunction was granted on behalf of Boulton and Watt and against the Ding Dong adventurers preventing the use of the Ding Dong engine.[29] In fine Boulton and Watt form Thomas Wilson backed up the injunction with threats:

> "...... you will therefore consider whether, it will be more your interest to accede to them (*Boulton & Watts terms*), Stop the Mine, or run the risk of incurring the Chancellors displeasure by continuing to work on."[30]

As a result of this injunction the Ding Dong adventurers had, by 27th August 1795, stopped the engine which meant that the mine could not be worked. Boulton and Watt rather tersely informed the adventurers that if they agreed to pay all the legal costs they had incurred and to enter into a regular agreement Boulton and Watt would "no doubt" grant them a licence to use the engine.[31]

Using Trevithick in an attempt to sidestep the 22nd March injunction understandably infuriated Boulton and Watt. Ambrose Weston, on behalf of Boulton and Watt, was anxious to establish Trevithick's association with Bull and to prevent him acting in concert with Bull in future:

> "Mr Trevethic Jun(io)r could not be included (*in the Ding Dong injunction*) as

he had not been made a party to the Suit, but if you will send me his Christian name, I will take care to prevent his future intermeddling: and further if you can send me an affadavit that Trevethic is Bull's partner or his known agent, I will apply to the Court to grant an Attachment against Bull for his contempt."[32]

In September 1795 an injunction was drawn up against Richard Trevithick Junior in regard to his activities at Ding Dong, although Trevithick appears to have been particularly adept at avoiding the bailiffs.[33] It took until December 1796 to serve the injunction when Trevithick was caught unawares in a public house adjacent to Boulton and Watt's Soho Foundry. A delighted Matthew Boulton commented that Trevithick seemed "much agitated and Vexed".[34]

Boulton and Watt obviously felt that, in spite of the ambiguous 1795 decision, the 1793 decision regarding Bull's piracy of the patent left them in an unequivocal position. This is certainly reflected in the number of injunctions Boulton and Watt issued with regard to Bull's engines. The threat of legal action seems to have significantly impacted on Bull's practice in the period after 1794. Up to the 22nd March 1794 injunction against Bull he had erected, although not necessarily completed, at least ten engines to his design but after that date possibly as few as three of Bull's engines were commissioned by Cornish adventurers.

The first mine to commission Bull to erect a new engine in the "post 22nd March injunction" period was Poldice. At the time the Boulton and Watt v Bull was being heard in Common Pleas in early 1795 the Poldice adventurers, led by R. A. Daniell, were in dispute with Boulton and Watt (who incidentally were also Poldice adventurers) due to arrears in engine payments. In what looks like an act of direct defiance to Boulton and Watt the Poldice adventurers decided to erect a Bull engine on the mine.[35] Both Edward Bull and Richard Trevithick Junior had a hand in producing the drawings for the Poldice Engine.[36] By the 7th March Bull had ordered the castings for the engine from Coalbrookdale, these had been completed by 8th July 1795.[37] In consequence an injunction was served on 13th July 1795 against Poldice and on behalf of Boulton and Watt.[38] In response to the injunction the Poldice adventurers "determined to stop all the Engines except Opies", effectively stopping the mine.[39] The stoppage of the Poldice engines triggered rioting on the mine at the end of July 1795, the miners rage being focussed on Boulton and Watt. Boulton and Watt regarded the riots as "an additional proof of the rascality of Bull's conduct and of the weakness of his intellect". In order to placate the riotous miners James Watt Junior suggested the miners should be informed that Boulton and Watt "have no objection to Bull's Engine being finished, provided our dues are paid, and Bull discharged from all interference". He also recommended "distributing a Sum of Money among the necessitous Miners, say £100, in order to counteract the pois(on) that has been circulated among them".[40]

That Bull's practice was failing was not lost on Trevithick who was increasingly developing his own independently of Bull. In the summer of 1796 Trevithick was attempting to come to terms with Boulton and Watt. In July 1796 Thomas Gundry wrote to Boulton and Watt on Trevithick's behalf:

> "I have taken the liberty to trouble you with this by desire of Rich(ard) Trevithick J(unio)r who have for some time past been employed by Edw(ard) Bull in Mechanism, he desires not to continue in opposition to you, and is ready to give up everything in this Countey and be under your direction........ if this step is taken I think the opposition in Cornwall would in a great measure subside."[41]

Trevithick's approach appears to have been initially, if cautiously, welcomed; however Murdock was opposed to the idea and it was soon dropped.[42]

On about 18th April 1797 Bull and Trevithick started a 63″ engine "on Bull's plan" at Wheal Treasure. This engine came second hand from Poldice. It is possible that the engine may have been a rebuild of Bull's 1793 63″ utilising parts from the Poldice 63″ of 1780 which had been stopped in August 1795.[43]

In May 1797 Edward Bull and Richard Trevithick Junior were talking to United Mines (which also counted R. A. Daniell amongst its adventurers) in regard to erecting one of their engines.[44] Trevithick approached Boulton and Watt asking for their consent to erect an engine at United which was roundly rejected, largely on the grounds that United was significantly in arrears with their payments to Boulton and Watt. Trevithick and Bull appear to have been attempting an attack on two fronts; Trevithick approaching Boulton and Watt formally whilst Bull threatened to act *ultra vires* if legitimate approaches were spurned. Boulton and Watt were contemptuous of Bull's stance, commenting on 2nd June 1797:

> "As to Bull's threats we know the Rascal is too well tied up by the Injunction to dare to move a step & the moment he breaks down the fence, we shall take care to have the scoundrel secured and the Adventurers put under a similar restriction."[45]

In spite of Boulton and Watt's blustering, Bull and Trevithick did erect an engine on United Mines. An affidavit sworn by William Murdock notes that in regard to the separate condenser and the air vessel the United engine was "exactly similar to the engine erected by Bull at Balcoath Mine".[46] On 31st January 1798 Ambrose Weston wrote to Wilson informing him that he was "preparing to file a Bill for an Injunction against United Mines in respect to their New Engine".[47]

Probably the last engine Edward Bull was directly involved in was erected at Pednandrea Mine at Redruth. On the 7th June 1797 Boulton and Watt offered to erect an engine on mine for "£1000 down or £55 per Month & the erection of the

Engine by Murdock".[48] This offer evidently did not find favour with the Pednandrea adventurers. They approached Edward Bull who was involved with the mine by late September 1797, if not earlier.[49] The engine was completed by November 1797. Trevithick's connection with the engine appears to have been limited to giving the adventurers some advice regarding the engine's valves after it had been started.[50] In an attempt to evade Boulton and Watt's patent Bull employed "various devices" on the Pednandrea engine:

> "It is to work with the power of the atmosphere, & instead of the air pump and condenser he has a preponderating column of water 33 feet long which forms his vacuum. Having got it ready to work Mr. Bull's Philosophy failed him for after he had been trying her to work 16 or 17 days successively to little or no purpose & the puzzled engineer had not been in bed for 13 days and nights......."[51]

Unfortunately Bull could not get the engine to work and, consequently, he was dismissed on Friday 17th November 1797. Bull was replaced by William Murdock who "extricated them from their difficulties by remedying the defects of the Engine" on the understanding that the Pednandrea adventurers would meet Boulton and Watts terms.[52]

The United and Pednandrea engines were to be the last erected by Edward Bull who died in March 1798 "of a putrid sore throat" (probably diphtheria) and was buried at Kenwyn Parish Church.[53] One wonders if his experiences with the Pednandrea engine contributed to his early demise?

Table 5: Engines erected by Edward Bull 1792 – 1797

Mine	Date	Cylinder	Notes	References
1. Balcoath	1792	20″	To Wheal Leeds, started 1st March 1794	Tann J., 1981; Tann J. 1986; Wilson papers
2. Wheal Rose	1793	24″	Syphonic engine To Wheal Billiard	Tann J., 1981; Tann J., 1986; Wilson papers
3. Wheal Treasure	1793	45″	Syphonic engine	Tann J., 1981; Tann J., 1986; Wilson papers
4. Wheal Treasure	May 1793	63″	Syphonic engine This may have been rebuilt in 1797 using parts from the B&W Poldice 63″	Tann J. 1996; Wilson papers
5. Carzize Wood	1793	30″	Syphonic engine Started 15th October 1793	Tann J., 1981; Wilson papers
6. Herland	1793	60″		Tann J., 1981; Wilson papers

7. Wheal Ann	1794	28″		Tann J., 1981; Wilson papers
8. Wheal Crenver	1793 / 1794	63″	February 8th 1794 (might have started May 1793)	Tann J., 1981; Wilson papers
9. Wheal Leeds	1794	20″	Started 1st March 1794	Tann J., 1981; Wilson papers
10. Hallamaning (a.k.a. Retallack)	1794	45″	(1) Under construction 22nd March 1794, started 1st April 1794, stopped end of June 1795 (2) To Wheal Abraham by 27th October 1796	Tann J., 1981; Tann J., 1996; Wilson papers
11. Ding Dong	Started 1795		Under construction 22nd March 1794, completed by Trevithick	Tann J.,1981; Wilson papers
12. Poldice	1795		Possibly not completed by Bull	Wilson papers
13. United Mines	1797			Wilson papers
14. Pednandrea	1797		Commenced by Bull, completed by Murdock	Wilson papers

Chapter 15 References

1. Dickinson H. W & Jenkins R., 1927
2. Wilson papers: AD1583/3/57
3. Wilson papers: AD1583/3/95 & 98, AD1583/11/69
4. Wilson papers: AD1583/3/102
5. Wilson papers: X208/24
6. *Ibid*
7. Wilson papers: AD1583/4/83
8. Wilson papers: AD1583/4/91
9. Wilson papers: AD1583/5/50
10. Wilson papers: AD1583/5/49
11. Tann., J., 1996; Wilson papers AD1583/11/69
12. Wilson papers: AD1583/6/31
13. Wilson papers: AD1583/8/54
14. Wilson papers: AD1583/5/64 & AD1583/6/5
15. Wilson papers: AD1583/6/5
16. Wilson papers: AD1583/6/18
17. Wilson papers: AD1583/6/21 & AD1583/6/24
18. Davies J., 1816; Dickinson H. W. & Jenkins R., 1927; Farey J., 1827; Muirhead J. H., 1859
19. Wilson papers: AD1583/7/15
20. Wilson papers: AD1583/7/19
21. Wilson papers: AD1583/7/22

22. Wilson papers: AD1583/7/22
23. Wilson papers: AD1583/7/26
24. Wilson papers: AD1583/7/56 & AD1583/8/9 & 10
25. Davies J., 1816; Dickinson H. W. & Jenkins R., 1927; Farey J., 1827; Muirhead J. H., 1859
26. Dickinson H. W. & Jenkins R., 1927; Webster T., 1844; Wilson papers: AD1583/8/63
27. Joseph, P. & Williams, G., 2015; Wilson papers; AD1583/8/40
28. Wilson papers: AD1583/8/37 & 40
29. Joseph, P. & Williams, G., Wilson papers: AD1583/8/27
30. Wilson papers: AD1583/8/34
31. Wilson papers: AD1583/8/75
32. Wilson papers: AD1583/8/27
33. Wilson papers: AD1583/8/74 & 89
34. Wilson papers AD 1583/9/54
35. Wilson papers: AD1583/8/1, 2, 4, 6, 16
36. Wilson papers: AD 1583/8/65
37. Wilson papers AD1583/8/49 & 52
38. Wilson papers: AD1583/8/60
39. Wilson papers: ADS1583/8/69
40. Wilson papers AD1583/8/71, 72
41. Wilson papers: 1583/9/37/2
42. Wilson papers: AD1583/9/38& 39
43. Wilson papers: AD1583/10/8 & 34, AD1583/11/69
44. Wilson papers: AD1583/9/76
45. Wilson papers: AD 1583/9/80
46. Wilson papers: AD 1583/10/12
47. Wilson papers: AD1583/10/9
48. Wilson papers: AD1583/9/83
49. Wilson papers: AD1583/9/101/1
50. Wilson papers: AD1583/9/112
51. Cited in Tann J., 1981
52. Wilson papers: AD1583/10/45 & 68
53. Harris T. R., 1977

Appendix 1
The men who built the engines.

"There are very good engine smiths in Cornwall with some bad ones (all of them love drinking too much)". Jonathan Hornblower (I) to James Watt.

John Budge (1733 – 1823)

A Camborne man, he has been described by Allen Buckley as the "foremost engine erector and steam engineer in the principal mining district of Cornwall by the 1770s".[1] Budge received his training at Dolcoath under George John. Buckley quotes a letter dated 14th December 1766 concerning Budge's appointment to erect "a new fire- engine not less than 40 – inches at Bullen Garden". This engine would be the first he erected out with the supervision of George John. Already noted is Budge's association with John Wise. On November 5th, 1772 John Budge was granted patent No. 1025 for "A machine for raising metals, minerals, or other heavy materials from great depths".[2] By 1793 Budge was the proprietor of the Tuckingmill Foundry. In addition to the Dolcoath engines, Budge appears to have erected a number of other atmospheric engines including those at Wheal Chance.[3] In 1777 John Edwards wrote to James Watt:

> "Though perhaps you may not find Mr Budge inclined to talk very scientifically, even on the subject of engines, yet almost half the engines in this county are under his care, and were built by his direction, and he is esteemed one of the best practical engineers in it."[4]

James Watt inspected five of Budge's engines in 1777; in a letter written to Matthew Boulton on 25th August 1777 Watt appears to have been be unimpressed:

> "I have seen five of Bonze's (Author's note: *Budge's*) engines..... but was far from seeing the wonders promised. They were 60, 63, and 70 inch cylinders. At Dalcoath and Wheal Chance they are said to use each about 130 bushels of coals in the 24 hours, and to make about 6 or 7 strokes per minute, the strokes being under 6 feet each. They are burdende to 6, 6½ and 7 lbs. per inch. One of the 60 inches threw out two cubic feet of hot water per stroke, heated from 60° to 165°. The 63 inches, with a 5 feet stroke, threw out 1½ cubic foot, heated from 60° to 159°, and so on with the others."[5]

Initially Budge appears to have been equally sceptical about Watt's engines. After a visit to the Soho foundry to see one of Watt's engines working he declined a commission to erect a Watt engine at Wheal Union (Tregurtha Downs). However the two men appear to have become reconciled. In 1779 Watt was very complementary about a Watt engine erected by Budge at Wheal Chance.[6]

Edward "Ned" Bull (c.1759 – 1798)
(See Chapter 15). Originally an engine erector for Boulton and Watt, by the early 1790s Ned Bull had branched out as an engineer in his own right, erecting engines which infringed Watt's patent. With the support of many Cornish adventurers, Bull was central to the legal challenges to Watt's patent. Edward Bull was Richard Trevithick's mentor and teacher during the mid 1790s and, as such his influence was felt throughout the nineteenth century. In a letter written on 16th February 1859 Matthew Loam (1794 – 1875) summed up Bull's career in a few short lines:

> "Mr Bull, I believe was a clever Man. But not to be compared, to Mr Murduct (Author's note: *Murdock*) but Bull, got jelous of Murduct and left Mr Watt, and set himself up in opposition, thought to evade the Patent right, by turning the cylinder upside down, and do away with the Air pump but the Engine was a failure, and poor Bull died."[7]

Bull's name came to be synonymous for inverted cylinder engines, giving Ned Bull a degree of immortality. A 70″ "Bull engine", built by Harvey & Co. in 1856, may occasionally be seen in steam at the London Museum of Water and Steam (Kew Bridge).

The Costers
John Coster (II) (1647 – 1718)
"The father of Cornish copper mining". He was initially involved in copper smelting in Bristol and the Wye Valley and, by the mid 1690s he was actively involved in Cornish mining with the backing of the Bristol Company. Coster pioneered both the driving of long adits and the use of large overshot waterwheels. In 1714 he patented a water engine with his son John (III).[8]

John Coster (III) (1687/8 - 1731)
Son of the above, he was living in Cornwall by 1714 to manage the family's mining interests. He was co-patentee with John Coster (II) of the 1714 water engine. It would have been largely down to his influence that Newcomen engines were erected in the county in 1725 – 7.[9]

The Hornblowers
The name Hornblower is intimately linked with engine building in eighteenth century

Cornwall. Joseph Hornblower erected some of the earliest engines in the county and founded a veritable dynasty of engineers. Separating the various Hornblowers from one another can at times be rather confusing as it was a family tradition that all Christian names would start with a J. Joseph's son Jonathan (I) was at the forefront of engine building in Cornwall in the post drawback period and was responsible for introducing Boulton and Watt to the county. Jonathan's sons were no less important: Jonathan (II) developed the compound engine under what must has been very trying circumstances whilst Jabez Carter was central to the legal challenges to Boulton and Watt's virtual monopoly on engine building.

Joseph Hornblower (1696 – 1761)

Joseph Hornblower erected Newcomen engines at North Downs, Chacewater, and Polgooth in 1725 – 7. Joseph hailed from the Midlands and it was in the Midlands that Joseph would have initially encountered Thomas Newcomen. The Baptist connection between the two men has been considered in Chapter 5. There is a tradition that Hornblower worked in Cornwall with Newcomen and it is possible that he may have gained experience working on the Wheal Vor Newcomen engine. Joseph became an erector of Newcomen engines in his own right, working initially in Shropshire before coming down to Cornwall in about 1725.[10] Cyrus Redding notes that after completing the Polgooth engine Joseph left the county entirely".[11] Whilst the 1725 – 1727 engines appear to have been the only ones erected by Joseph in Cornwall he did erect further engines elsewhere in England and Wales, purchasing iron cylinders from the Coalbrookdale Company.[12] Joseph was the first of what may be described as a dynasty of engine erectors and steam engineers.

Jonathan Hornblower (I) (1717 – 1780)

The first of Joseph's descendents to make their mark in Cornwall was Jonathan Hornblower (I). Jonathan, like his father, was a Midlander, having been born in Broseley on 30th October 1717. Also like his father Jonathan was a Baptist, in 1767 he was largely responsible for establishing a Baptist meeting house in Chacewater He appears to have learnt his trade from his father, assisting him to erect engines in Derbyshire, Shropshire and Wales. Jonathan moved to Cornwall in 1745 initially establishing himself at Truro, latterly living at Chacewater. Cyrus Redding, in *Yesterday and today* of 1863, notes that his grandfather, Jonathan Hornblower (I) erected his first Cornish Newcomen engine on Wheal Virgin. However given that work at Huel Virgin did not commence until 1757 it is somewhat difficult to reconcile this with Jonathan's arrival in Cornwall in the autumn of 1745, although this does correspond with an increase of engine building in the county.[13] He established himself as a successful engineer erecting engines on mines including Wheal Virgin, Wheal Sparnon and Tresavean. John Smeaton noted that, alongside John Nancarrow, Jonathan Hornblower was one of the principal engineers in the County.[14] Jonathan had a reputation as a strong man, his grandson Cyrus Redding recounts the following story:

"One day he heard some of the men disputing at Huel Virgin about their strength – and Cornishmen who work above ground are not weak men, as their wrestling shows. He bade them take up a fifty pound weight which lay at hand, and one after another to see how far they could throw it. When they had tried, he took it up and threw it farther than any of them."[15]

On his arrival in Cornwall in 1777 James Watt described Jonathan Hornblower as "a very pleasant old Presbyterian". Initially Watt seems to have viewed Jonathan Hornblower as something of a sceptic in regard to the "new faith" noting on August 14th 1777 that Jonathan "seems a good sort of man and carries himself very fair but is I hear an unbelieving Thomas".[16] This seems somewhat unjust given that the first Boulton and Watt engine ordered in Cornwall, in November 1776, was for Ting Tang; Jonathan Hornblower being the engineer for the mine. On his deathbed his dying words were reputed to have been occasioned by the numerous relatives gathered around his bed: "Oh take me away from all these who are come to see a mortal die."[17]

Jabez Carter Hornblower (1744 – 1814)
Born at Broseley, Jabez was the eldest of Jonathan's (I) sons. In 1765 he was earning 12s 6d a week assisting his father erecting a Newcomen engine at Wheal Sparnon. In 1775 Jabez was employed building engines for the Dutch Government. By 1778 he was back in Cornwall assisting his father at Poldice. With the transition from the Newcomen engine to the Boulton and Watt Jabez found employment with Boulton and Watt, as an engine erector earning a guinea a week. At some point Jabez fell foul of his employers and was dismissed. He later worked with Jonathan Hornblower (II) although by 1790 he was collaborating with J. A Maberley on an engine which infringed Watt's patent. It was this engine which formed the basis of the legal actions of 1796 and 1798/9 which proved, in law the validity of Watt's patent.[18]

Jethro Hornblower (1746 – 1820)
Erected engines for his brother Jonathan Hornblower (II) during the 1790s.

Jonathan Hornblower (II) (1753 – 1815)
(See Chapter 14) Inventor of the compound engine which he patented in 1781. Building engines in Cornwall in the 1790s. Sadly Jonathan Hornblower's career was severely constrained by Boulton and Watt. If Hornblower had been able to develop his engine without hindrance and had managed to extend his patent it is highly probable that Jonathan would have pre-empted Woolf's work on the compound engine. Given the clear run that he deserved Jonathan Hornblower (II) would have made a much more significant contribution to the development of the Cornish pumping engine.

For those wishing for further details on the Hornblower clan the following article is recommended: Harris T. R, (1976), The Hornblower family, Pioneer steam engineers, *Journal of the Trevithick Society*, No. 4, 1976, pp. 7 – 44.

William Murdock (1754 – 1839)

Boulton and Watt's chief erector in Cornwall, Murdock entered their service in 1777. In September 1779 Murdock came to Cornwall on Boulton and Watt's behalf. William Pole paints an (uncritical) portrait of a man who was a skilled engineer and a diplomat of no little ability:

> "In a history of the Cornish engine, it would be unjust to omit the mention of Mr. William Murdock, who was one of Messrs. Boulton and Watt's principal assistants, and for sixteen years previous to the expiration of their patent resided at Redruth, as manager of the engines they had erected in the district. He was the author of many valuable minor inventions in the details of the engine, and it was in great measure owing to his skill and ability that the machine so rapidly attained its high degree of perfection. Mr Watt himself bears testimony to his ingenuity and skill, to which, he says, were due many improvements in the manufacture of engines, and in the machinery and tools used for that purpose.
>
> Mr Murdock was always exceedingly popular in Cornwall, and has left behind him a character which all must admire; much of the success of Watt's engines there being attributed to his great capabilities and to his able and discreet management. During the unhappy disputes which for so long kept Watt and the Cornishmen at variance with each other, and amidst all the ill feeling which these dissensions engendered, Murdock's unflinching integrity, combined with his amiability of disposition, retained for him unchanged the esteem and respect of both parties. His only ground of disagree-

William Murdock.

ment with the Cornish managers was that they sometimes would not allow him to spend money enough to execute their work well. His zeal for the improvement and efficient working of the engine was indefatigable; it is said of him, "that he has spoiled many a good coat" by going into flues and other dirty places, to satisfy himself by personal observation as to the causes of derangement, &c. or for experimental enquiries necessary in the course of improvement."[19]

Murdock left Cornwall in 1798, returning to Birmingham where he spent the rest of his life. He remained with Boulton and Watt until 1830, as a partner from 1810. To be both respected by the Cornish and valued by James Watt was no mean feat. He was a talented engineer and inventor, whose interests included the use of steam for vehicle propulsion and the use of gas for lighting and yet his career is one of unfulfilled potential. This seems to have due to a combination of natural modesty and creditable loyalty to Boulton and Watt. This latter was, some would say shamelessly, played upon by his employers, not always to Murdock's advantage.

William's given name was *Murdoch* but, on coming south, he 'anglicised' it to Murdock and thus it appears in contemporary sources. Perhaps he wearied of spelling it to Sassenachs. The form *Murdock* has therefore been used throughout this book.

John Nancarrow

In spite of John Smeaton's description of John Nancarrow as one of the principal engineers in Cornwall little is known of the man or his work. He was mine Captain at Wheal "Rith" and also was responsible for overseeing the engines at Great Work. According to the Great Work cost book Nancarrow in 1764 was being paid £4 a month plus £1 1s for inspecting the engines. In 1764 he was paid £30 "for making the new boiler and inspecting the engine".[20] Nancarrow, along with William Lemon, was Borlase's informant regarding Newcomen engines. John Nancarrow may have been a very early exponent of the expansive use of steam and, by implication, higher working pressures. John's son, also John, emigrated to America where he became a celebrated engineer in his own right.[21]

Tom Pearson

One of Boulton and Watt's erectors. Like Murdock, Pearson was a Scot. He was described by Matthew Loam (1794 – 1875) as "a droll Fellow, was something like Paddey (*an Irishman*), always in Love, or in Liquor". Much of Pearson's spare time was spent consorting with one Gracie Carnsew of Crowan, known to her contemporaries as "Ungracey" she was, by all accounts, a woman not afraid to lift the elbow. Gracie, whilst in her cups, was in the habit of abusing Edward Bull, one of her ditties was recorded for posterity:

"My dear Tom Pearson, thee dost know about an Engine, thee dost know about the cross on the Bottom of the Cylinder, but what do that chucklehead Bull know, he don't know anything about it....."[22]

Quite what Edward Bull had done to deserve Gracie's scorn, history does not record. After leaving Cornwall, Pearson "was lost sight of" although latterly he secured a position working the engines at Crofton on the Kennet and Avon Canal.

Richard Trevithick Junior (1771 – 1833)
In the nineteenth century Trevithick would gain enduring fame as the father of the steam locomotive and a pioneer of high pressure steam; however the importance of his activities during the latter years of the eighteenth century have been over represented by his son Francis in his 1872 biography. During the mid 1790s Trevithick was learning his trade whilst working in a subsidiary role with Edward Bull. Towards the end of the century Trevithick started to develop a practice as an engineer in his own right, proving to be a minor thorn in the side of Boulton and Watt.

Sampson Swaine
Swaine is chiefly remembered for Patent No. 774 of May 21st, 1762 which covered the use of waste heat from smelting to heat a boiler and a device to derive rotary motion from a Newcomen engine. Swaine is believed to have been a St Agnes man, although he moved to Camborne around 1730. In the 1740s Swaine established a copper smelter on Rosewarne Downs at Camborne, the furnace having been designed by Swaine.[23] A further copper smelter was established at Entral. Pryce notes that:

> "about the year 1754, ... one Sampson Swaine, in conjunction with some gentlemen of Camborne, erected furnaces at Entral in that parish; but their situation being too remote from coal, they removed their works to Hayle.[24]

The relocation to Hayle took place in 1758, the company becoming the Cornish Copper Company, the only large-scale copper smelter in Cornwall.

Thomas Wilson (1748 – 1820)
Thomas Wilson was not an engine builder; however as Boulton and Watt's Cornish agent his role was critical to their business in the county. Wilson was a Yorkshireman who moved to Cornwall in 1775 to take up the agent's post at Chacewater Mine.[25] Presumably Boulton and Watt's involvement on the mine led to his appointment as their agent in Cornwall. Given the sometimes strained relationship between Boulton and Watt and the Cornish adventurers Wilson must have been a diplomat of no little ability. In return for his services he was paid a commission of 2½ per cent of the engine premiums. Between 1781 and 1800 Wilson received £3,485 in commission. Beyond his relationship with Boulton and Watt Wilson had numerous business interests in the county including mine adventuring, farming, shipping, pack mules, the supply of candles and brewing. Wilson died at the age of 72 in 1820 and was buried in Falmouth.[26] The survival of Boulton and Watt's correspondence

with Wilson is absolutely fundamental to our understanding of Boulton and Watt's activities in Cornwall.

John Wise (1697 – 1775)

Like Joseph Hornblower John Wise did not confine his activities to Cornwall, having also erected engines in his home county of Warwickshire and also in Bristol and London. He was active as an engine builder from the 1720s to the 1770s. On October 21st 1740 Wise was granted Patent No. 571 for converting the Newcomen engine's reciprocating motion to a rotary motion:

> "The fire engine itself works with its latest improvements, but instead of drawing or forcing water with pumps, there is a fixed chain rope, or rod to the end of the working beam, which comes down perpendicular to my new invention or machine, which is under a separate roof from the fire – engine, where is a horizontal shaft on which are wheels moveing (that is to say,) one wheel fixed on the shaft with iron rounds, on which run iron dogs with teeth, to catch in the said rounds to move the said shaft half round, and then quit the same; the second is a double tumbling wheel, to which the above chain and dogs are fixed, and is also moved half round with a shaft by the perpendicular stroke from the fire – engine, and then returns back againe to its place by a weight, whilst a fly keeps the said shaft in its proper motion to make it circular.
>
> But if it is required for large heavy works such as plating, rolling of copper, iron, and so forth, then it works by throwing the raised water through cylinders or trunks on an overshott wheel fixed to the horizontal shaft, which will, with a spring or stream of water of two hogsheads by the hour, do the same work as other watermills that require one thousand hogsheads by the hour."[27]

John and his wife Elizabeth moved to Cornwall in around 1750 where he was employed at Dolcoath working latterly in conjunction with John Budge. Wise continued to work at Dolcoath until 1771 when his name ceases to appear in the pay records.[28]

For a well researched account of John Wise's activities readers are recommended to seek out the following article: Grudgings S., (2012b), John Wise – Unrecognized engine builder and contemporary of Newcomen and Watt, *International Journal for the history of engineering and technology*, vol. 82, No. 2, July 2012, pp. 176 - 186.

Appendix 1 References

1. Buckley A., 2010
2. Anon, 1872
3. Cited in Dickenson H. W. & Rhys Jenkins, 1927
4. Cited in Smiles S., 1865
5. Dickenson H. W. & Rhys Jenkins, 1927

6. McKay D., 2010
7. Stewart R. J., 2015a
8. *Ibid*
9. Allen & Rolt, 1997, Harris T. R., 1976
10. Redding C., 1863
11. Harris T. R, 1976, Raistrick A., 1953
12. Borlase W., 1758
13. Farey J., 1827, Harris T. R., 1976
14. Redding C., 1863
15. Cited in Dickenson H. W. & Rhys Jenkins, 1927
16. Redding C., 1863
17. Harris T. R., 1976
18. Pole W., 1844
19. Borlase W. 1758, Harris T. W., 1977
20. Harris T. W., 1977
21. McKay D., 2010
22. Buckley A., 2010; Carter C., 1998
23. Pryce W., 1778
24. Wilson papers: AD1583/11/68
25. Cited in Dickinson H. W. & Jenkin R., 1927
26. Anon, 1871
27. Buckley A., 2010, Grudgings S., 2012

Appendix 2
Pumps

Throughout this book the focus has been almost entirely on engines, little attention being paid to the pumps that the engines were driving; this appendix is intended to go some way to remedying that inbalance.

A—Sump. B—Pipes. C—Flooring. D—Trunk. E—Perforations of trunk. F—Valve. G—Spout. H—Piston-rod. I—Hand-bar of piston. K—Shoe. L—Disc with round openings. M—Disc with oval openings. N—Cover. O—This man is boring logs and making them into pipes. P—Borer with auger. Q—Wider borer.

Woodcut from Agricola's *De Re Metallica* of 1556 showing the boring of timber pump bodies, a lift pump in use (note the clack valve F) and the pump stripped to its components. This technology would have been very familiar to Cornish Miners in the first quarter of the eighteenth century. *Image courtesy Tony Clarke.*

At the time the Costers made their appearance in the south west, lift or bucket pumps (also referred to as sucking pumps) were the order of the day. This remained the dominant technology throughout most of the eighteenth century. This was not new technology; Agricola in his *De Re Metallica* of 1556 illustrates this type of pump. At the beginning of the eighteenth century the pump barrel or body would typically comprise bored out lengths of timber. Due to its dimensional stability when either wet or dry, elm was a popular choice of timber for pump bodies.

A lift or bucket pump may be thought of as comprising two sections: the bottom section where water is drawn up the pump column by suction and the upper section of the column where water is lifted by a piston or "bucket". The two sections of the column being separated by a clack valve which will allow water to pass through it from the bottom section of the pump but will not allow water to pass from the upper section down into the bottom section. The piston or bucket also operates as a one way valve, on a descending stroke water will pass through the piston but not on a rising stroke. The piston is connected to the engine (water or fire) by means of wooden pump rods.

On a upward stroke the piston will lift the water in the upper section of the pump discharging it either into a cistern or into an adit. The action of the rising piston will also draw water up the bottom section of the column from the sump or a cistern in which the bottom of the pump sits. The water from the lower section of pump will pass through the clack valve and into the upper section. On a downward stroke the piston will descend in the column and the cycle starts again. The amount of water lifted on each stroke will be dependent on the stroke of the piston and the diameter of the pipe.

Multiple pumps could be used in combination draining into and drawing out of shared cisterns. When pumps were used in combination they were referred to as a lift or tyer / tier of pumps (confusingly either term can be used to refer to a series of pumps being worked in sequence or a single pump within that sequence). Pryce notes:

> "Tyer – or Tier of Pumps. A set of pumps belonging to the engine, of which the lower pump or piece is called the Driggoe, but more frequently the Working - piece; the other have names appropriated to them, as the Tye or Adit –lift, the Rose – lift, the Crown – lift, the Lilly, the Puppy &c. each being a separate Tier of Tyer."[1]

As the eighteenth century progressed and engineering improved, there was an increasing use of cast components, gradually replacing both wood and parts fabricated by blacksmiths. Whether this innovation is attributable to the Costers is not known, however, given their very close links with the Bristol brass industry, they would surely have availed themselves fully of the benefits of this improved technology. In his will John Coster II left his west of England mining interests to his wife Mary and his sons Thomas and John III including:

> " ..... all my Water Engines Tin or Iron Bellmettle Cyllinders and Brasses and all other Engine Materials...."[2]

Certainly by the 1720s cast items are being recorded as being employed in the manufacture of mine pumps. Robert Bald in his *General view of the coal trade of Scotland* of 1808 contains an "Account of the Expenses of Edmonstone Fire – Engine, to Mr Potter, discharged 1st July 1727". The account is directly contemporary with engines erected in Cornwall by Joseph Hornblower in 1725 – 1727 and so is of particular value and interest. The engine was supplied completed with wooden pumps:

> "Paid for elm pumps at London £53 4s 0d
> To two cast – mettle barrels, 9 foot long, and 9 inches diameter, and with expenses after them, £41 16s 6d
> To two brass buckets, two clacks, 9 inches diameter."[3]

From the account one can infer that there were two pumps, the bottom pump lifting water from a sump to a cistern, the second lifting from the cistern to a discharge point either at surface or an adit. The pump bodies were constructed from elm. The "cast mettle – barrels, 9 foot long, and 9 inches diameter" would be used to replace the bottom of the upper section of each column. This section of the pump is known as the working barrel. For the pump to work efficiently the piston must fit flush in the column requiring very accurate boring of a wooden column for the length of the stroke of the piston. Replacing this section of the wooden column with a cast cylinder, allows much more accurate fitting of the piston. Likewise the piston or buckets and clacks are also cast allowing greater accuracy than items which would previously been fabricated by a blacksmith.

The increasing use of cast parts in pump construction during the first half of the eighteenth century reflects the improvements in foundry technology being made during the period. This is mirrored in the development of engine cylinders. In the 1710s brass casting technology was sufficiently advanced to allow the casting of engine cylinders. Brass, however, was an expensive commodity and by the early 1720s foundry technology had reached a point where cast iron started to replace brass in engine manufacture. Thus the 1720s saw a transition from brass to iron cylinders, Coalbrookdale starting to cast iron cylinders in 1722.[4]

Given the improvements in cast iron technology it comes as no surprise that cast iron replaced wood in the manufacture of pump barrels. Whilst Coalbrookdale would certainly have been able to supply pump barrels in the 1720s, the demand appears to have been limited prior to 1733. After that date demand increases, typical orders being between 100 and 345 feet of pipe. For example in 1734 Coalbrookdale supplied Joseph Hornblower with 345 feet of pump pipes. As demand for pump barrels increased their production became more remunerative to Coalbrookdale than the supply of engine cylinders.[5] One can safely assume that the increase in the demand for iron pump pipes in the 1730s marked the transition from wood to iron pump barrels. If so it is probable that all the post drawback engines in Cornwall were using cast pump barrels. However wood pump barrels were not wholly ousted from Cornwall during the eighteenth century: In 1781 Cooks Kitchen paid Richard Bennetts and partners sixteen shillings for boring a pump.[6] Björling in his *Pumps*

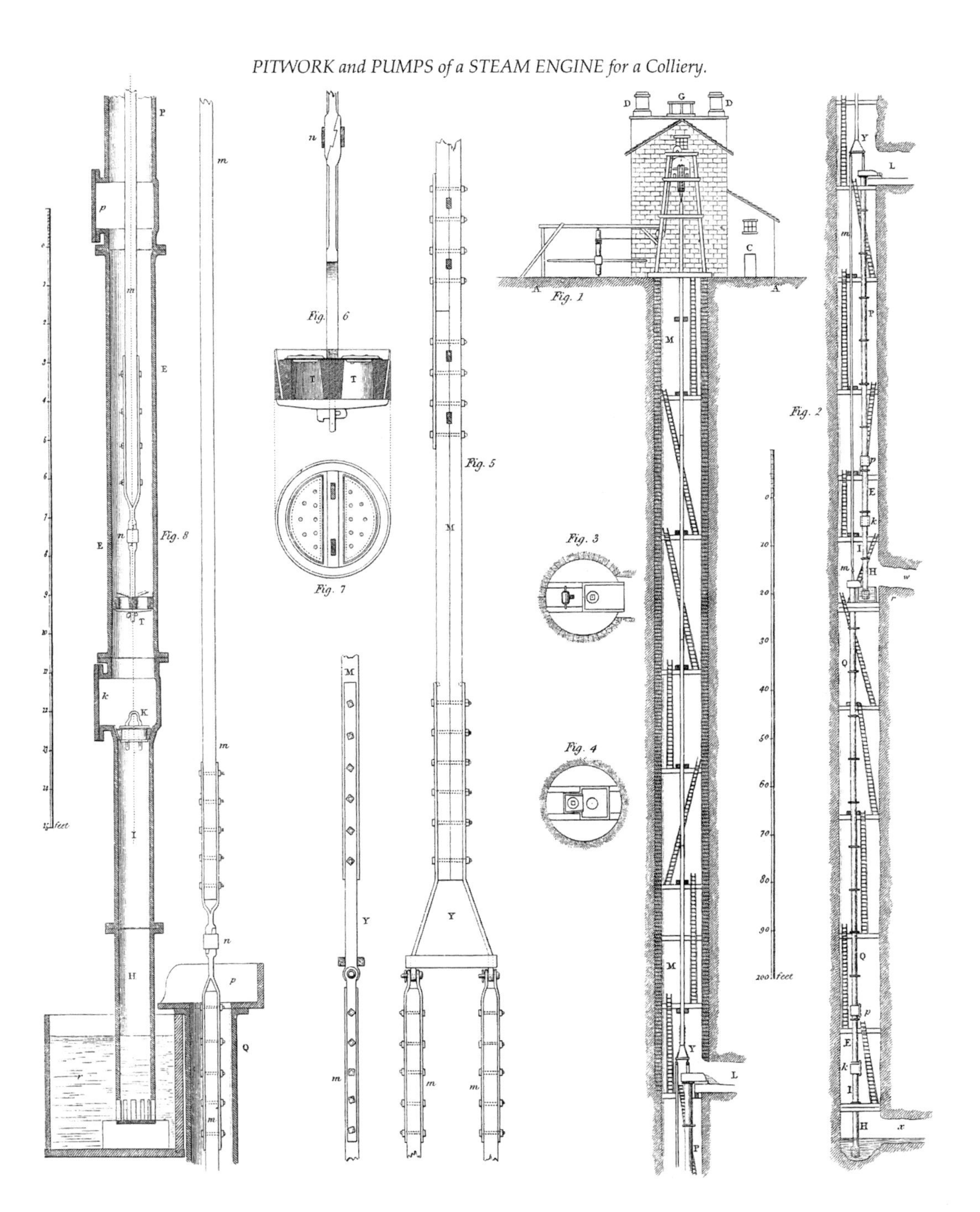

Illustration from Farey's *Treatise on the Steam Engine* of 1827 showing the arrangements of lift pumps at Long Benton Colliery. The arrangement shown would have been typical of what would have been found on the majority of Cornish mines during the nineteenth century. *Image courtesy Tony Clarke.*

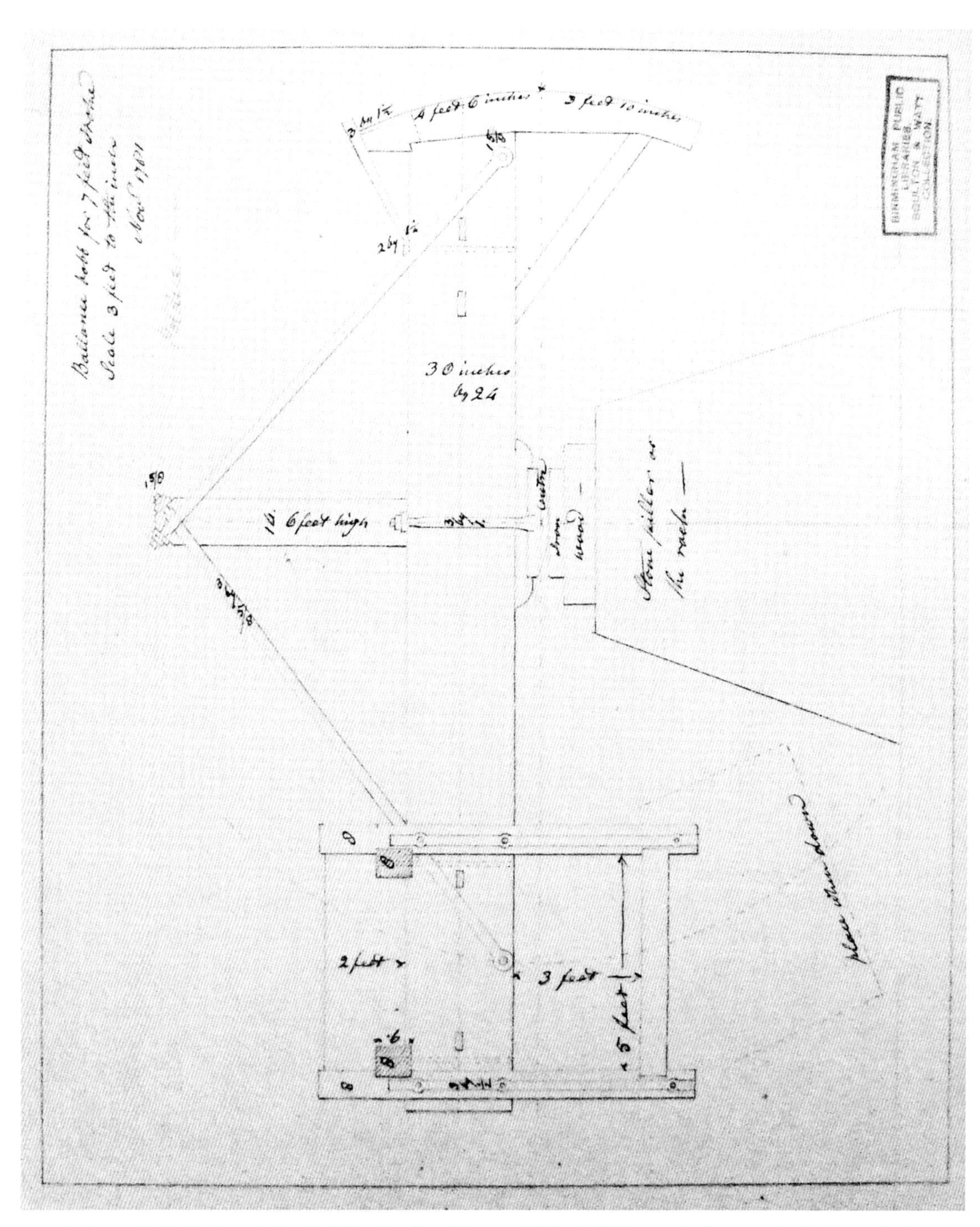

Balance bob designed for Poldice by Boulton and Watt. Balance bobs counterbalanced the weight of pump rods in deep shafts. Reproduced with the permission of the Library of Birmingham, MS 3147/1286.

historically, theoretically and practically considered comments: "In the year 1798, we still found them boring out the wood pumps in Cornwall".[7]

When used in conjunction with a single acting steam engine this type of pump had an inherent flaw. The power stroke of the engine had to lift both the weight of the water in the pump column and also the weight of the wooden pump rods. This was not such an issue with the water engines of the period which typically drove two parallel sets of pumps via cranks at ninety degrees to each other, thus the weight of a falling pump rod in one lift would, in combination with the water engine's bobs, counterbalance the weight of both pump rods and water in the other lift. Having only one set of pumps meant that the steam engine did not have the inherent counterbalancing advantage of the water engine operating two lifts of pumps. To counterbalance the weight of rods and water most steam engines working at any sort of depth utilised a balance bob.

Apart from reducing the load on the engine a balance bob would also help to reduce shock loading when the pump rods reached the bottom of the return stroke. If the return stroke of the engine was too fast it could cause significant damage to both engine and pumps. As mines grew increasingly deeper, pump rods grew longer and heavier exacerbating the problem. To reduce the weight of the pump rods necessitated a rethink regarding the arrangement of rods in a shaft. In a traditional Newcomen engine each separate pump would have its own pump rod running right to the top of the shaft, being connected to the arch head by a chain. This was recognised as a problem and addressed by Smeaton by the time the Long Benton engine was erected in 1772. Rather than running each rod up beyond the top pump Smeaton directed that a single pump rod should be used to link the pumps to the arch head. Just above the top pump a single pump rod would terminate in an iron triangle known at the Y as it resembled an inverted Y. From each arm of the inverted Y hung a pump rod, one operating the top pump, the other continuing down the shaft to operate the second pump, further Ys could be used as required. In a deep shaft this would significantly reduce the weight an engine had to lift on its power stroke with a consequent improvement in pumping efficiency.[8] Smeaton's use of the inverted Y appears to have been an intermediate stage in the development of pump rods. Latterly a large single pump rod would be employed, running from the top of the shaft down to the bottom pump with short subsidiary rods to operate intermediate pumps as required, this arrangement would become dominant during the nineteenth century when the plunger pump eclipsed the bucket or lift pump.

Although the plunger or force pump is really part of the nineteenth century story rather than the eighteenth its use in Cornwall can be traced from the last years of the eighteenth century. The plunger pump was by no means a new invention, Pole observes that it was invented and patented by Sir Samuel Morland in 1675.[9] Unlike a lift or bucket pump which lifts water on the rising stroke of the pump rods the plunger pump lifts the water on the falling stroke of the rods by actuating a plunger which forces the water up the column in much the same way as a plunger forces water out of a syringe. In the lift or bucket pump the engine has to lift both the weight of the pump rod and the weight of the water whereas with a plunger pump the engine only has to lift the weight of the rods, the water being forced up the pump column by the weight of the falling rods on the engine's (non powered) return stroke. This obviously has hugely significant implications for pumping efficiency.

Boulton and Watt were using plunger pumps on a limited scale in Cornwall as early as 1786; on 6th March 1786 James Watt wrote to Thomas Wilson:

> "I am very Glad to hear that Wheal Messa behaves so well long may it continue. I also think it a very Good scheme to work the upper lift by the weight of the rods using a forcing pump."[10]

Similarly of the 17th September 1786 Matthew Boulton wrote to Watt regarding an order for a large double acting engine for Wheal Virgin:

> "Captn. Paul (& indeed every one of the Captns.) are very fond of the plungers or forcing pumps. 1st because they save balle bobs totally with their friction & vis inertia. 2. they save ¾ of the leather. 3. save time & stoppages as they never go out of order. 4. saves the ware of the pumps from the iron rods wch go down them. 5. saves the iron rods themselves. 6. saves much of the engine power as they evidently work quicker, and from all I can learn of the pitmen & everybody about the engines it is certainly better than balle bobs."[11]

William Pole notes:

> "Among other things, Murdock proposed the use of Sir Samuel Morland's plunger pump into the pit – work. Not as a general substitute for, but as an addition to, the lifting pumps, in order to suit the double acting engine, by making the pumps double – acting also. In 1796, one of these was employed at *Ale and Cakes*, a mine now forming the eastern part of United Mines.
>
> When Captain Lean was manager of Crenver and Oatfield mines, to which he was appointed in 1801, he there first introduced the use of the plunger pump to *supersede* the lifting pumps, wherever he found it practicable. This alteration was found so advantageous, that it was ultimately brought into general use, and is now become a system in the Cornish mines, the whole of the pumps in the shaft being of the plunger description, except the lower lift........ is still made a lifting pump."[12]

Appendix 2 References

1. Pryce W., 1778
2. John Coster's will 6th Oct 1716, PCC PROB 11/566; www.bittonfamilies.com

3. Bald R., 1808
4. Mott R. A., 1964
5. *Ibid*
6. Harris T. R., 1966
7. Björling P. R., 1895
8. Farey, J. 1827, Enys J. S. 1862
9. Pole W. 1844
10. Wilson papers; AD1583/1/86
11. Cited in Dickinson H. W. & Jenkin R., 1927
12. Pole W., 1844

Bibliography

Agricola G, (1556), *De Re Metallica,* (Hoover translation).

Allen J. S. & Rolt L. T. C., (1997), *The steam engine of Thomas Newcomen,* Landmark Publishing Ltd.

Anon, (1794), *The repertory of arts and manufactures*: Vol. 1, London.

Anon,(1839), Thomas Savery, *The Gentleman's Magazine*, Vol. XII, July – December 1839, p. 261.

Anon, (1867), *Abridgements of the specifications relating to preparation and combustion of fuel,* Patent Office.

Anon, (1868), Cornish Families. The Lemons, *One and all: A Cornish monthly Illustrated Journal, Newsletter and Record of Local History,* August 1868, Penzance.

Anon (1871), *Abridgements of specifications relating to the steam engine A.D 1618 – 1859,* Vol. 1, Patent Office, London

Bagnall J. N., (1854), *A History of Wednesbury in the County of Stafford,* William Park, Wolverhampton.

Bald R., (1808), *A general view of the coal trade of Scotland,* A. Neill & Co., Edinburgh.

Barton D. B., (1961), *A history of copper mining in Cornwall and Devon,* D. Bradford Barton Ltd., Truro.

Barton D. B., (1964), *A historical survey of the mines and mineral railways of East Cornwall and West Devon,* D. Bradford Barton Ltd., Truro.

Barton D. B., (1966), *The Cornish Beam Engine*, D. Bradford Barton Ltd., Truro.

Berg T. & P., (2001), *R. R. Angerstein's illustrated travel diary 1753 - 1755*, Science Museum.

Björling P. R., (1895), *Pumps historically, theoretically and practically considered,* 2nd edition, E. & F. Spon

Borlase W, MS Letter Books, Morrab Library, Penzance.

Borlase W., (1758), *The natural history of Cornwall*, Oxford.

Brooke J., (1993), *The Cunnack manuscript*, The Trevithick Society, Camborne.

Brooke J., (1996), Wheal Fortune in Ludgvan, *Journal of the Trevithick Society*, No. 23, 1996, pp. 63 – 67.

Brooke J., 2001, *The Kalmeter Journal*, Twelveheads Press.

Buchanan B. J., (2000), The Africa trade and the Bristol gunpowder industry, *Transactions of the Bristol & Gloucestershire Archaeological Society*, No. 118, 2000, pp. 133 – 156.

Buckley A, (2016), *The Great County Adit*, The Trevithick Society, Camborne.

Buckley A., (2010), *Dolcoath Mine – a history*, The Trevithick Society, Camborne.

Burt R., 1969, *Cornish mining, essays on the organisation of Cornish mines and Cornish mining economy*, David & Charles, Newton Abbot.

Carew R., (1602), *Survey of Cornwall*

Carne J., (1828), On the period of commencement of copper mining in Cornwall: and on the improvements which have been made in mining, *Transactions of the Royal Geological Society of Cornwall*. Vol. 3 1828, pp. 35 – 85.

Carter C., (1998), Early engineers around Camborne. *Journal of the Trevithick Society*, No. 25, 1998, pp. 52 – 64.

Claughton P., F. (1994), Silver – lead – A restricted resource: technological choice in the Devon Mines, *Bulletin of the Peak District Mines Historical Society*, Vol. 12, No. 3, summer 1994, pp. 54 – 59.

Claughton P. F., (1996), The Lumburn Leat – evidence of new pumping technology at Bere Ferrers in the 15th century, *Mining History: The Bulletin of the Peak District Mines Historical Society* Vol. 13, No. 2, Winter 1996

Cletscher T. (1696), *Relation of the European Mines in the year of 1696*. MSS (*this document may date from 1698 and not 1696*).

Davies J., (1816), *A collection of the most important cases respecting patents of invention and the rights of patentees*, W. Reed, London.

Day J., (1977), The Costers: copper-smelters and manufacturers, *Transactions of the Newcomen Society*, Vol. 47, 1974 – 1976, pp 47 – 58.

Desaguliers J. T, (1744), *A Course of Experimental Philosophy,* Vol. 2, London.

Dickinson H. W & Rhys Jenkins, (1927), *James Watt and the steam engine* (2nd Ed 1981), Moorland Publishing.

Durnford C. & Hyde East E., (1827), *Term reports in the Court of the King's Bench: Vol. III, Containing from Michaelmas Term, 39 George III. 1798, to Trinity Term, 40 George III. 1800*, H. C. Carey & I. Lea, Philadelphia.

Earl B., (1978), *Cornish explosives*, The Trevithick Society, Camborne.

Enys J. S., (1862), Remarks on the duty of the steam engines employed in the mines of Cornwall at different periods, *Transactions of the Institution of Civil Engineers,* Vol. III, pp. 449 - 466

Farey J., (1827), *A treatise on the steam engine, historical, practical and descriptive*, Longman, Rees, Orme, Brown and Green.

Farey J., (1971), *A treatise on the steam engine, historical, practical and descriptive* Vol. 2, David & Charles, Newton Abbot.

Gilbert D., (1830), On the progressive improvements made in the efficiency of the steam engines in Cornwall, *Philosophical Transactions of the Royal Society*, Vol.

120, pp. 121 – 132.

Greener J., (2015), Thomas Newcomen and his Great Work, *Journal of the Trevithick Society*, No.42, 2015, pp. 63 – 126.

Gregory O., (1806), *Treatise of mechanics, theoretical, practical and descriptive*, Vol. 2, George Kearsley, London.

Grudgings S., (2012a), Jarrit Smith's 1751 Newcomen engine, *South Gloucestershire Mines Research Group.*

Grudgings S., (2012b), John Wise – Unrecognized engine builder and contemporary of Newcomen and Watt, *International Journal for the history of engineering and technology*, vol. 82, No. 2, July 2012, pp.176 - 186.

Grudgings S., (2015), From Calley to Curr: the development of the Newcomen engine in the eighteenth century, *Proceedings of the NAMHO conference 2014, Welsh Mines Society, pp. 31 – 46.*

Hamilton J., (1854), *Excelsior: Helps to progress in religion, science and literature*, Vol. 2, James Nisbet and Co., London.

Harris T. R., (1945), Engineering in Cornwall before 1775, *Transactions of the Newcomen Society*, Vol. XXV, 1945 – 47 pp. 111 – 122.

Harris T. R., (1966), *Arthur Woolf: The Cornish engineer*, D. Bradford Barton Ltd, Truro.

Harris T. R, (1976), The Hornblower family, Pioneer steam engineers, *Journal of the Trevithick Society*, No. 4, 1976, pp. 7 - 44

Harris T. R., (1977), Some lesser known Cornish Engineers, *Journal of the Trevithick Society*, No. 5, 1977 pp. 27 – 65.

Hatchett, C., (1967) *The Hatchett Diary*, 1796, edited by A. Raistrick, D. Bradfgrd Bartn, Trur.

Hills R. L., (1997), James Watt in Cornwall, *Journal of the Trevithick Society*, No. 24, 1997, pp. 46 – 60.

Hornblower J. C., (1801), Account of a fact respecting water heated in a boiler of stone, with observations, *A Journal of Natural Philosophy, Chemistry and the Arts*, Vol. 3, pp. 169 – 171.

Howard B., (1999), The duty on coal 1698 – 1831, *Journal of the Trevithick Society*, No. 26, 1999. pp. 30 – 35.

Jars G, (1823), *Voyages metallurgiques ou recherches et observations sur les mines, les forges, etc*. Tome 3, Paris

Jenkin A. K. H., (1927), *The Cornish Miner*, George Allen & Unwin Ltd., London.

Joseph, P., (2012). *So Very Foolish, A History of the Wherry Mine, Penzance*, The Trevithick Society.

Joseph P. & Williams, G. (2015). *Ding Dong Mine, A History*. The Trevithick Society,

Camborne.

Lemon C. (1838), Statistics of the copper mines of Cornwall, *Transactions of the Statistical Society of England* June 1838. (Reprinted in Burt R., (1969), *Cornish Mining*, David & Charles, Newton Abbot.

London Gazette issue 5459, 11th August 1716, p 2.

The Lord Hawkesbury, (1799), *Report from the committee appointed to inquire into the state of the copper mine and copper trade of this kingdom. 7th May 1799.*

McKay D (transcriber), (2010), A letter from Matthew Loam (1794 – 1875) to his nephew Matthew Loam (1819 – 1902), *Journal of the Trevithick Society*, No. 26, 1999. pp. 95 – 109.

Messenger M., (2015) *Caradon & Looe, the canal, railway and mines*, 3rd Edition, Twelveheads Press, Chacewater.

Mott R. A., (1964), The Newcomen engine in the eighteenth century, *Transactions of the Newcomen Society,* Vol. XXXV, 1962 – 1963, pp. 69 – 86.

Muirhead J. P., (1854), *The origin and progress of the mechanical inventions of James Watt,* Vol. 2, John Murray, London.

Muirhead J. P., (1859), *The life of James Watt with selections from his correspondence*, 2nd Edition, John Murray, London.

Pennington R. R., (1975), The Cornish Beam engine and patent law, *Journal of the Trevithick Society,* No. 3, pp. 45 – 57.

Pennington R. R., (1977), The Cornish Metal Company, 1785 – 1792, *Journal of the Trevithick Society,* No. 5, pp. 76 – 88.

Pole W., (1844), *A Treatise on the Cornish Pumping Engine*, John Weale, London.

Polwhele R., (1831), *Biographical Sketches in Cornwall Vol. 1,* London.

Pryce W., (1778), *Mineralogia Cornubiensis,* London; reprinted D. Bradford Barton Ltd., Truro.

Raistrick A., (1953), *Dynasty of Ironfounders, The Darbys and Coalbrookdale*, Longmans, Green & Co. Ltd., London.

Redding C., (1863), *Yesterday and Today*, Vol. 1, Newby.

Rhys Jenkins, (1942), The Copper Works at Redbrook and Bristol, *Transactions of the Bristol and Gloucestershire Archaeological Society*, Vol. 63, pp 145 – 167.

Rogers K. H, (1976), *The Newcomen engine in the West of England*, Moonraker Press, Bradford-on-Avon.

Rowe J., (1993), *Cornwall in the age of the industrial revolution*, 2nd edition, Cornish Hillside Publications, St Austell.

Savery T., (1702), *The Miners Friend or an engine to raise water by fire described* (1827 reprint), London.

Skempton A. W. (Ed.), (1981), *John Smeaton FRS*, Thomas Telford Ltd.

Smeaton J, (1759), An experimental enquiry concerning the natural powers of water and wind to turn mills and other machines depending on a circular motion, *Philosophical Transactions of the Royal Society* Vol ? pp 100 – 174.

Smeaton J, (1812a), Some remarks concerning the design for a double boring machine for cylinders and guns to be erected upon the tail – water of the Carron Works, *Reports of the late John Smeaton*, Vol. 1, pp. 276 – 279.

Smeaton J, (1812b), Chase Water Engine, *Reports of the late John Smeaton*, Vol. 2, pp. 347 – 359.

Smeaton J., (1837), *Reports of the Late John Smeaton F.R.S,* Vol. 2. M. Taylor.

Smiles S., (1865), *Boulton and Watt, comprising also a history of the invention and introduction of the steam engine,* John Murray.

Smiles S., (1866), *Boulton and Watt, comprising also a history of the invention and introduction of the steam engine 2nd ed.*, John Murray.

Stewart R. J., (2015a), The Coster family and its contribution to mine drainage in south west England, *Journal of the Trevithick Society*, No. 42, 2015, pp. 127 – 150.

Stewart R. J., (2015b), Deep mining technology: Devon & Cornwall in the eighteenth century, *Proceedings of the NAMHO conference 2014*, Welsh Mines Society, pp. 19 – 30.

Stewart R. J., (in preparation), *John Smeaton and the fire engine 1765 – 1785*. (Paper for 1st International Early Engines Conference, 2017)

Tann J., (1981), Mr Hornblower and his crew: Watt engine pirates at the end of the eighteenth century, *Transactions of the Newcomen Society*, Vol. 51, 1979 – 80 pp. 95 – 109.

Tann J., (1996), Riches from copper: The adoption of the Boulton & Watt engine by Cornish mine adventurers, *Transactions of the Newcomen Society*, Vol. 67, 1995 – 96 pp. 27 – 51.

Thomas R., (1819), *Report on a survey of the mining district of Cornwall, from Chasewater to Camborne*, John Cary, London; reprinted D. Bradford Barton Ltd., Truro.

Todd A. C., (1960) and Jonathan Hornblower (1753 – 1815), *Transactions of the Newcomen Society*, Vol. XXXII, 1959 – 60 pp. 1 – 13

Todd A. C., (1967), *Beyond the blaze: A biography of Davies Gilbert*, D. Bradford Barton Ltd., Truro

Tonkin MSS, Royal Institution of Cornwall collection.

Torrens H. S., (1982), New light on the Hornblower and Winwood compound steam engine,_*Journal of the Trevithick Society*, No. 9, 1982, pp. 21 – 41.

Torrens H. S., (1984), Some newly discovered letters from Jonathan Hornblower (1753 – 1815),_*Transactions of the Newcomen Society*, Vol. 54, 1982 - 83 pp. 189 – 200.

Trevithick F., (1872), *The life of Richard Trevithick*, Vol. 1, Spon, London.

Watson J. Y., (1843), *A Compendium of British Mining with Statistical Notices of the Principal Mines in Cornwall,* London. Reprinted D. Bradfrd Bartn, Trur.

Wilson Papers. Online transcriptions at www.cornishmining.net/story/bwpapers.htm (originals held at the Cornwall Record Office, Truro).

Wilson T., (1792), *A comparative statement of the effects of Messrs. Boulton and Watt's steam engines with Newcomen's and Mr Hornblower's*, W. Harry, Truro.

Wilson T., (1844), *Reports and notes of cases on letters patent for inventions*, Thomas Blenkarn, London.

Index

Recent publications from the Trevithick Society.

Despite their significant place in the story of Cornish mining, the mines of the Great Flat Lode have never received the level of attention accorded to those situated north of Carn Brea. Mining historian Allen Buckley has now remedied a part of this omission in his most welcome coverage of the history Basset Mines. His book begins with fascinating insights into the earliest days of mining in the Carnkie area, progresses through the boom years of the nineteenth century, ending with the valiant struggle for survival of Basset Mines Limited, one of the largest mines ever to work in Cornwall, until its closure in 1918. As well as a mass of historical information, the book also contains a gazetteer of surviving sites, which will be of great value to those who seek to explore an area of industrial archaeology, equal in interest to any in Cornwall.

Working almost continuously between 1810 and 1877, this fabulously rich mine kept going during periods of low tin prices which led to the failure of many other mines. It is also reputed to be the first Cornish mine to install a Newcomen steam pumping engine. One of only three Cornish mines to have its own smelting works, it produced up to a quarter of Cornwall's tin from a main lode of unparalleled richness. The 1850s saw an attempt to pump out and rework the mine, using the largest ever installed in Cornwall. This venture incurred the greatest loss ever in a single attempt to re-open a mine. This comprehensive history by Tony Bennett puts Wheal Vor in its rightful place amongst the county's greatest mines.

This book recounts the history of the Cornish fuse works. It details William Bickford's invention of the miners' safety fuse in 1831; the subsequent history of Bickford, Smith and Company's factory in Tuckingmill is detailed, together with descriptions and histories of the other fuse works in Camborne, Redruth and elsewhere. The book contains maps and many old and new photographs, including some Bickford-Smith family photographs published here for the first time.